AF556257

Growing Strawberries

R.R. Sharma
M.Sc., Ph.D. (Hort.)
Scientist
Division of Fruits and Horticultural Technology,
Indian Agricultural Research Institute, New Delhi-110 012

INTERNATIONAL BOOK DISTRIBUTING CO.

Published by

INTERNATIONAL BOOK DISTRIBUTING CO.

(Publishing Division)
Chaman Studio Building, 2nd Floor,
Charbagh, Lucknow 226 004 U.P. (INDIA)
Tel. : Off. : 450004, 450007, 459058 Fax : 0522-458629
E-Mail : ibdco@sancharnet.in

First Edition. 2002

ISBN 81-85860-89-0

Composed & Designed at :

Panacea Computers
33, Nehru Road,
Sadar Cantt. Lucknow-226 002
Phone : 481164, 480546
E-mail : prasgupt@rediffmail.com

Printed at:

Army Printing Press
33, Nehru Road,
Sadar Cantt. Lucknow-226 002
Phone : 481164, 480546
E-mail : armypress@sify.com

GROWING STRAWBERRIES

Affectionately dedicated to

my reverend mother

Smt. Vidya Sharma

Foreword

Strawberry is one of the most delicious and refreshing fruits of the world, which is eaten not only as a source of nutrition but also for its sweet sour taste and pleasant flavour. It is a commercial fruit of temperate regions but with the development of day neutral varieties, its cultivation can be undertaken in tropical and subtropical climates. In India, it was first introduced in early sixties but efforts to boost its cultivation got set back because of poor adaptability of the introduced cultivars and poor knowledge on technical know how about its cultivation. However, later day neutral cultivars were introduced and agro-techniques were standardized, which led to phenomenal increase in the farmers growing strawberries. Initially, its cultivation was confined only to hilly tracts of India but with the introduction of Floridian cultivars like Chandler, Douglas, Fern etc. its cultivation spread to tropical and sub-tropical zones also. Now, a number of farmers are growing strawberry commercially in many parts of our country.

However, it is really very unfortunate that strawberry even being a highly remunerative crop, there is hardly any literature available in India on its cultivation. Thus, it is a dire need of the hour to have literature on such a valuable crop. I am happy that Dr. R.R. Sharma, a young scientist has come up with the idea of compiling the available literature on strawberry in the form of a book entitled "Growing Strawberries". He has compiled the latest available information and has put it systematically in different chapters. This book covers all aspects of strawberry cultivation in India and abroad. Special emphasis has been given on propagation, nursery management, plant protection and physiological disorders. In my opinion, it is a complete treatise on package of practices of strawberry cultivation.

I am sure that the book will attain great popularity among farmers, students, teachers, researchers, and extension workers, dealing in the field of fruit growing.

P.K. Singh

(P.K. Singh)
Director
Indian Agricultural Research Institute
New Delhi - 110 012

Place : IARI, New Delhi -12
Date : January 21, 2002

Foreword

[illegible] is one of the [illegible] of the world, which is known not only as a source of nutrition but also [illegible]

[illegible]

[illegible] continued only [illegible] but with the introduction of [illegible]

[illegible]

[illegible] Dr. H.P. Singh [illegible]

[illegible]

I strongly feel that the book will attain great popularity among [illegible] students, teachers, researchers and extension workers dealing [illegible]

[illegible]
Director

Place: New Delhi [illegible] Indian Agricultural Research Institute
Date: [illegible] New Delhi - 110012

Preface

In India, all types of fruits are grown with variable intensity because India is blessed with almost all types of climates. It ranges from extreme dry temperate of Ladakh to extreme tropical climate of South India. This is the reason that India, at present, is the second largest fruit producing country of the world after China. Strawberry is grown commercially in temperate regions of many countries of the world. Initially, breeding efforts were made only to develop short-day cultivars, which were most successful in temperate climate. However, later, a fashion of day-neutral cultivars originated, which started boosting its cultivation in warmer localities also.

In India, many cultivars of strawberry were introduced in the hilly tracts during early sixties to popularize them but could not be made popular because the introduced cultivars had poor adaptability in the hills. Moreover, there was no standardized agro-technique about its cultivation. However, later, day-neutral cultivars were introduced and accordingly different agro-techniques were standardized. It resulted in boost in its cultivation. In the recent years, farmers have shown much more interest in its cultivation.

Despite being such an important fruit crop, there is no proper literature available on Strawberry cultivation and if at all some information is available then that is not compiled and written in comprehensible language. Realizing the acute problem, I have attempted to compile the available information in the form of a book.

This book is written in a simple and understandable language. It covers almost all information on different aspects of strawberry cultivation practiced in different parts of the world, including India. It is hoped that this book will be of a great help to researchers, teachers, extension workers, students and alike in the field of strawberry cultivation.

I acknowledge with thanks the help and inspiration rendered by

Dr. Room Singh, Head, Dr. A.M. Goswami, (Ex-Head) Dr. V.P. Sharma, Sr. Scientist, Dr. Sanjay Kumar Singh, Scientist, and several students of Division of Fruits and Horticultural Technology, IARI, New Delhi, for their time to time help during the preparation of the manuscript. I am also thankful to Mr. Surinder Kumar for taking beautiful photographs, Dr. (Mrs.) Karuna Dixit for drawing beautiful sketches and M/s Kushwaha Computers, Inderpuri, New Delhi, for typing the manuscript.

In the last but not least, I am especially thankful to my wife, Veena Sharma, whose *filial* attachment did not come in my way during the preparation of this manuscript.

R.R. Sharma

New Delhi - 110 012

CONTENTS

Chapter **Page**

Chapter 1
Introduction

The strawberry is known as the most delicious and refreshing fruit of the world to millions of peoples. Its plant is also cherished in gardens and commercial plantations for its beautiful and attractive red fruits. Besides, its fruit is a rich source of vitamins and minerals with delicate flavour. Thus, strawberry has become an important table fruit of millions of people throughout the world. The added advantage with strawberry is that it gives early and very high returns per unit area compared to other fruits because its crop is ready for harvesting within six months after planting. Moreover, its fruits remain available as fresh throughout the year.

It can be grown in wide climatic conditions, ranging from temperate to tropical climate. Since its cultivation is greatly influenced by specific regional adaptation due to critical photoperiod and temperature requirement, its cultural practices are highly variable. But extensive research, breeding high yielding cultivars, standardization of agro-techniques and adoption of its scientific cultivation by the farmers have boosted its production tremendously, as a result of which, it has got widespread popularity in the recent years.

Nutritive value and uses

The fresh, ripe fruit of strawberry is a rich source of vitamins and minerals (Table 1). It is fairly a good source of vitamin A (60 IU/100 g of edible portion) and vitamin C (30-120 mg/100 g of edible portion). Strawberry is also a rich source of pectin (0.55 per cent) which is available in the form of calcium pectate and serves as an excellent ingredient for jelly making. It is also a rich source of potassium, calcium and phosphorus, the total soluble solids (TSS)

Table 1 : Composition and nutritive value of strawberry fruit

Water (%)	85-90
Protein (g)	0.25-0.7
Carbohydrates (g)	8.5-9.2
Fats (g)	0.2-0.5
Fibre (%)	1.1
Sugars (%)	
Fructose	1.5-3.5
Glucose	1.5-3.0
Sucrose	0.8-2.5
Organic acids (mg/100 g of edible portion)	
Citric	400-1250
Malic	90-675
Succinic	90-100
Oxalic	20-24
Tartaric	15-17
Vitamins	
Vit. A	60.0-63.0 IU
Ascorbic acid	30-120 mg
Thiamine (B_1)	0.03 mg
Riboflavin (B_2)	0.07 mg
Niacin (B_6)	0.06 mg
Minerals (mg/100 g ofedible portion)	
Potassium	160-164
Calcium	20-21
Phosphorus	20-21
Iron	1-1.2
Sodium	1-1.2
Total phenolics	60-120 mg
Anthocyanins	50-100 mg

content being 7-12 per cent. Water is a major constituent (90 per cent) of strawberry fruit. The mature soft fruit contains about 5.0 per cent total sugars and 0.90 to 1.85 per cent acids. Fructose and glucose are major sugars in its fruit with a small proportion of sucrose. Citric acid is the most abundant organic acid in strawberry, followed by malic, succinic and oxalic acids. These acids determine the pH and colour stability, inhibiting the enzyme activity of the strawberry fruit.

The strawberry is known for its characteristic aroma. It is due to the presence of volatile esters. Important aroma compounds are ethyl hexanoate, methyl hexanoate, ethyl heptanoate, ethyl propionate, ethyl butanoate, methyl butanoate, furanone and linalool. However, concentration of these volatile compounds varies amongst cultivars and thus the aroma. It also contains lipids with appreciable quantity of oleic acid. Essential oil can also be extracted from its leaves. Linalool and nonanal are major constituents of strawberry oil.

Ripe strawberries attain attractive red colour on maturity and have soft melting flesh of a characteristic flavour. The red colour of fruit is mainly due to the presence of anthocyanin, pelarogonidin-3-monoglucoside and traces of cyanidin. The fully ripe strawberry fruit has a sweet sour taste with a pleasant aroma. The fruits are mostly eaten as fresh and consumed mostly not for their food value but for their flavour. In addition to dessert purpose, strawberries are also processed into various value-added products like canned strawberry, jam, jelly, ice-cream, conserves, freeze strawberry, wine, other soft drinks etc. The strawberry jam and strawberry flavoured icecream are popular worldwide and liked by one and all.

Chapter 2

Origin, History and Distribution

Origin of modern strawberry

The modern cultivated strawberry (*Fragaria x ananassa* Duch.) is a hybrid between two largely dioecious octoploid species, *Fragaria chiloensis* and *Fragaria virginiana*. However, some strawberry cultivars have also been derived from a subspecies of *F. virginiana* and *F. virginiana glauca*. It is assumed that hybridization of *F. chiloensis* and *F. virginiana* occurred spontaneously in Europe in the 1700s, when female plants of *Fragaria chiloensis* of Chilean origin were grown in the proximity of male *Fragaria virginiana* plants of North American origin. Since then, extensive hybridization between the parent species and their descendents had occurred, making *Fragaria x ananassa*, a highly variable and heterozygous species.

History and distribution

The history of strawberry goes back to the Romans and perhaps even the Greeks, but because the fruit has never been a staple of agriculture, it is difficult to find out an ancient reference to it. Theophrastus, Hippocrates, Dioscorides and Galen did not even mention it, nor did Cato, Varro, Columella or Palladius, the four world famous Latin writers on agriculture. Apulius cited the strawberry only for its medicinal value. Although Virgil and Ovid did name the strawberry in their verses but they did so only casually in poems of the country life, where they associated it with other wild fruits. However, it is evident from literature that by 1300's, strawberry was in cultivation in Europe and then the French began to transplant the wood strawberry (*Fragaria vesca*) from the wilderness to the garden. But the plant was considered more ornamental for its flowers than the use for its fruits, although it was grown to some extent for table purpose also. The first reference of strawberry cultivation dates back to 1368, when a gardener, Jean Dudoy, of King Charles V, planted 1,200 plants of strawberry in the royal gardens of the Louvre in Paris. In 1375, Chateau de Couvres planted four blocks of strawberries in the garden in Duchess of Burgundy. No doubt that its cultivation was crude but it was highly appreciated by the Duchess of Burgundy. England too was an early admirer of strawberry which is evident from the fact that the successive modifications of the name from Streowberige, Streowberge, Streberewyse and strawberry took place in England, showing a long familiarity with it there. There are many theories in England about the derivation of the name strawberry. Anglo-Saxons called the strawberry as "hayberry" because it ripens at the time the hay is mown. Another guess is that the name derived from the way children strung the berries on straws of the grass or hay to sell them, a custom still practiced in some parts of Ireland even today. A more likely explanation is that the Anglo-Saxons used the name strawberry to describe the way the runners strew or string away from the mother plant to find a place to grow.

The 15th century has horticultural significance for establishing strawberry as a common garden plant in France. The common garden strawberry in France was *F. vesca*, the wood strawberry. The other species of Fragaria like *F. moschata* (the musky flavoured strawberry), *F. virdis* (the green strawberry), *F. alba* and *F. sylvestris* were also described by Jerome Bock in 1532. But at the end of 15th century, two cultivated strawberries in the gardens were wood strawberry (F. *vesca*) and the musky strawberry, *F. moschata*, both characterized by small distinctly flavoured fruits. The only significance of the 1600's to strawberry history was the introduction of *F. virginiana* to Europe from eastern North America, for this berry was to be the sire of today's modern big-fruited strawberry. Though, the true story about the journey of Virginiana strawberry to Europe is still unknown.

However, a few workers suggest that it reached Europe in the early seventeenth century.

The word strawberry dominated in the late 1600's. The musky, the green and the Virginiana strawberries were grown to a minor extent. The first half of the next century brought the dramatic development of the commercial large-fruited strawberry. The importance of the seventeenth century had been the introduction of *F. virginiana*, the first step in bringing together the future parents of modern strawberry. As stated earlier, the Virginiana strawberry was the sire of the modern strawberry (*Fragaria ananassa*).

In 1712, a French army officer, A. F. Frezier, returned from Chile to France, where he had seen the large-fruited strawberry (*F. chiloensis*) at Concepcion and brought five plants with him. After reaching France, he distributed these plants to various persons. He gave two plants to Cargomaster of the ship, one plant to the king's garden in Paris, one to his superior of Brest and kept one with himself (Darrow, 1966). Thus, the journey of the Chilean strawberry *F. chiloensis* from Chile to France in 1712 was the most important event in the history of modern strawberry, because the Chilean strawberries had attractive large-sized berries, which attracted the gardeners there. By the mid of the seventeenth century, it was grown commercially. Its hybrids, which were accidental results of using *F. virginiana* as polliniser for *F. chiloensis,* were more fertile and adaptable. Some gardeners in England, Holland and France made careful selections for large-fruited varieties of *F. ananassa*, as the hybrid was called. Later, it travelled to different parts of the world. In India, a few plants of strawberry were brought in the early sixties by NBPGR Regional Station, Shimla, from where it spread to other states.

Chapter 3
Botany and Taxonomy

Botany

Strawberry plant is a low creeping, perennial herb in which stem is compressed into a resettled crown, with internodes of about two mm in length. The axillary buds in leaf nodes of the crown either remain dormant, or develop into branched crowns (Plate 1), or stolons (runners), depending on the prevailing environmental conditions. Inflorescence form terminally and vegetative growth is continued by the uppermost axillary bud of the crown, resulting in a sympodial growth habit. Although, it is often referred to as a herbaceous perennial, aging results in lignification of the crown, producing a hard woody tissue. With time, extensive branch crown development produces a structure resembling a highly compacted tree scaffold, with certain physiological responses to the environment similar to those of deciduous fruit trees. However, vegetative and reproductive growth of the strawberry are more sensitive to environmental conditions, particularly photoperiod and temperature, than most of the other fruit crops. The different parts of strawberry plant have been illustrated in Figure 1.

Root system : Different species and varieties have different sized root systems, depending largely on whether they make runners freely and express their vigour in number of runners per plant, or whether they make few runners and express their vigour in making large individual plants. Usually, strawberry has two types of roots, the large primary roots, which originate in crown and small secondary lateral roots that make up the mass of the root system, arising from the primary roots. There are usually 20-35 roots but may be up to

100 or more primary roots and thousands of small rootlets in a good root system. Generally, root development is rapid in the fall and spring, when there is less demand for water by the leaves. However, extensive root development may take place in December-January also. The wider adaptation of strawberry is mainly due to its variable root system, which is perennial only in part. In general, strawberry roots grow during one year and die the next, during the fruiting season but when all flower clusters are removed from a mature plant, most of its roots do not die at fruiting time. Roots of strawberry grow chiefly downward in well-drained sandy soils and a few roots may be found as deep as 60 cm. In clay soil, they spread more horizontally. In general, 90 per cent of roots occur in the top layers (25-30 cm) of the soil.

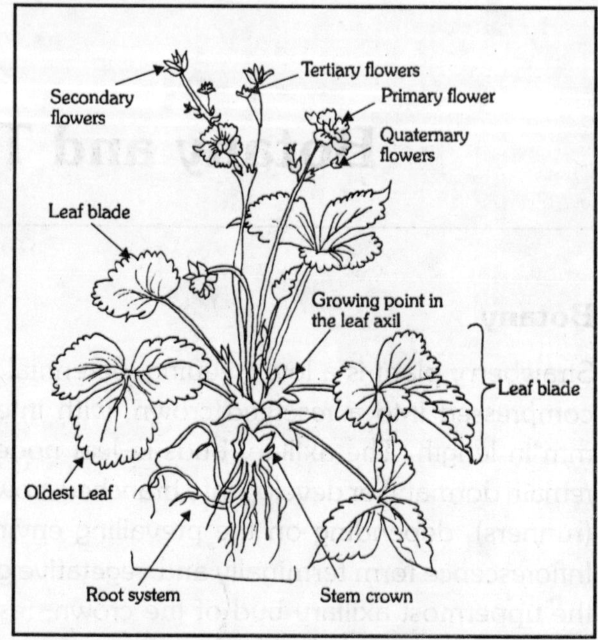

Figure 1. Strawberry plant

Leaf : The leaves are arranged in a 2/5 spiral, each sixth leaf being just above the first for maximum light exposure. The leaves of *F. vesca* are thin, those of *F. chiloensis* thick and of *F. virginiana* intermediate. The thin leaves of *F. vesca* are characteristics of the humid climate plant, while those of *F. chiloensis*, with thick cuticles and deep set stomata, are characteristic of a dryland plant.

Crown : The strawberry crown is actually the very shortened stem of the plant. Sometimes, it may become as much as 60 cm in length and nodes longer apart, as in sand dunes of Pacific Coast. During frost injury, its crown is the first part to be damaged.

Symbiotic association : Sometimes, a certain type of fungus is found in the roots of a strawberry plant. It has been found to be mycorrhizal in nature, i.e. the fungus may furnish nutrients to the host plant. The commonly found mycorrhizal fungi in strawberry plantations are species of *Rhizophagus* or *Phycomycetous*. But, in general, different species of strawberry have specific need for mycorrhizal fungi. However, there is every possibility of mycorrhizal fungi becoming parasitic under some conditions.

Growth habit : When a runner is set, new leaves appear with a bud in the axil of each leaf. During summer, some of the buds remain dormant, some develop into runners and occasionally one develops into a branched crown. Depending on variety and climatic conditions, the buds in leaf axils develop more often into a branched crown and flower buds. Runners are produced during summer from the buds in the axils of new leaves, and in succession as the leaves develop. The first axillary bud, which differentiates in the spring, becomes the first runner and so on. These runners have two nodes each and internodes of the same length. The more rapidly the plant grows, the more runners are produced. The size and the length of the runners usually depend on the growth conditions and varietal characteristics.

Taxonomy

Fragaria species belongs to family Rosaceae, with basic chromosome number of x = 7. The cultivated strawberry, *Fragaria x ananassa,* is an octoploid, having chromosome number (2n) of 56. In addition to *F. x ananassa*, the genus *Fragaria* includes at least 17 other species, including diploids, tetraploids, octoploids, and a hexaploid. Many of these species have been cultivated at one time or the other and some are still being grown though on a limited scale. Of the 17 species, F. *chiloensis, F. daltoniana, F. nilgerrensis* and *F. vesca,* grow in different parts of the world. On the basis of chromosome number, the different species have been classified into four different groups (Table 2).

Table 2 : Distribution of strawberry species and their ploidy level

Ploidy level & chromosome number	Major species	Distribution
2x = 14	*F. vesca, F. viridis* *F. nilgerrensis, F. daltoniana* *F. nubicola, F. iinumae* *F. yesoensis, F. nipponica* *F. mandshurica*	Central Asia and Far East
4x = 28	*F. moupinensis, F. orientalis,* *F. corymbosa*	East and South East Asia
6x = 48	*F. moschata*	Europe, North and South America
8x = 56	*F. chiloensis* *F. virginiana* *F. iturupensis* *F. x ananassa*	Central and North America, Chile, Peru, Kuril islands of Japan and Hawaii

Chapter 4

Species of genus *Fragaria*

Species

Till now, seventeen species of strawberry have been reported. These have been divided into four groups on the basis of chromosome number, with seven chromosomes as the base number (x). Of these, nine are diploids, three tetraploids, one hexaploid, and four octoploids. The cultivated strawberry, *Fragaria x ananassa,* is also an octoploid. A brief description of different species of *Fragaria* is given hereunder:

1. Diploids (2n = 2x = 14) : *Fragaria vesca* L. (wood strawberry), is the most extensively distributed of all the species *of Fragaria*. It is found in North America, northern Asia, northern Africa, and Europe. Its plants are erect and produce runners profusely. Plants have thin, light green leaves borne on slender petioles, producing inflorescence of the same length as that of leaf petioles. Flowers are bisexual and generally self-fruitful; fruits are long, ovate, bright red in colour, and usually highly aromatic. It has four subspecies, viz. *F. versca ssp. vesca* (Europe and Asia); ssp. *americana* (eastern North America to British Columbia); ssp. *bracteata* (western North America); and ssp. *californica* (California). All the subspecies are self-fertile except for *ssp. bracteata*, which bears both hermaphrodite and occasionally female flowers.

The *Fragaria viridis,* a native of Europe, is found along the edges of the forests. Its plants are slender and upright, producing few runners without nodes. The leaves are deep green. It bears small inflorescence, and bisexual flowers, larger than that of *F. vesca*. The fruits are small, firm, pink to red in colour, and aromatic. It produces a few viable seeds after self-pollination.

The *Fragaria nilgerrensis* is a native to south-east Asia and China. Its plants are robust and spreading, producing strong runners. The leaves are pubescent, dark green and heavily veined. The inflorescence is small, bearing bisexual flowers and small, nearly round, pale-pink and relatively tasteless fruit with many seeds.

The *Fragaria daltoniana* occurs in a small area of the Sikkim Himalayas (India) at higher altitudes (3,000-4,500 m). The plants are vigorous producing slender runners. The leaves are petiolate, shiny, coriaceaous with few indentations on the margins. The flowers are solitary and fruit may be elongate or fusiform, 2-2.5 cm long, bright red in colour and nearly tasteless.

The *Fragaria nubicola* is found in the then USSR and Sikkim in the Himalayas (India) at an altitudes of 1,500 to 4,000 m above mean sea level. The plants closely resemble those of *F. vesca* but are dioecious rather than hermaphrodite.

The *Fragaria iinumae* is found in the Alpine mountains of central and northern Japan. Its plants are vigorous and erect with slender filiform runners. The leaflets are shortly petiolate and coarsely dentato-serrate. Flowers often have 6-8 petals. The leaves are glaucous like *F. virginiana* ssp. *glauca*. Scape is one or few, erect, with 1 or 2 flowers. Fruit is ovoid, nearly tasteless with achenes embedded in pits on the fruit.

The *Fragaria yesoensis* is native to Japan. The plants are deciduous and completely dormant in winter. *Fragaria nipponica* is found in the mountains of Japan. It closely resembles to *F. yesoensis* but has pedicles with ascending hairs. The fruits are ovoid to sub-globose in shape, and have an unpleasant taste. The *Fragaria mandshurica* is a native of Manchuria. It is similar to the autotetraploid, *F. orientalis*, which may be its progenitor species. It largely produces hermaphroditic flowers but is self-infertile. The fruits are high in acid and have yellow achenes in shallow pits.

2. Tetraploids (2n = 4x = 28): The *Fragaria moupinensis* is one very important tetraploid species of *Fragaria*, found mainly in eastern Tibet, Yunnan and western China. Its plants are dioecious and closely resemble to *F. nilgerrensis*. The leaves are trifoliate, with smaller leaflets on the petiole below. The flower stems are longer than the leaf petioles, with 2 to 4 flowers/inflorescence. The fruits are small and orange-red in colour.

The *Fragaria orientalis* is chiefly found in western Siberia, Mongolia, Manchuria and Korea. The plants are small and upright, producing slender runners. The leaves are ovate, nearly sessile, and light green, with deeply serrated margins. The inflorescence has a few large flowers. The fruit is soft, conical to round, and slightly aromatic. The *Fragaria corymbosa* is probably a native to northern China, but has been described only recently and presently it is known that it has only male plants.

3. Hexaploids (2n = 6x = 42): The *Fragaria moschata* is the only hexaploid species of *Fragaria* which occurs in northern and central Europe, Russia and Siberia. The plants are dioecious, vigorous and tall, producing runners freely. The leaves are large, heavily veined, rugose, pubescent and dull green. The inflorescence is longer than the leaf petioles, bearing large flowers. The fruit is dark red, soft, irregular-globose to ovoid, with aromatic or musky flavour. The cultivated forms have perfect flowers.

4. Octoploids (2n = 8x = 56): The *Fragaria chiloensis* (L.) is the most important octoploid species of *Fragaria*, found along the Pacific coast of Alaska, central California, Chile and Hawaii. This species is now cultivated only to a limited extent, but was once extensively grown in Chile, Peru, and Ecuador. Its plants are low spreading and vigorous with prolific runner producing capacity. Wild populations of *F. chiloensis* are primarily dioecious, although hermaphrodites have been found in California and Chile. The leaves are thick, dark green and very glossy. The fruits are very large, dull to bright red, firm, with white flesh and pungent flavour. The *F. chiloensis* ssp. *chiloensis, F. chiloensis* ssp. *lucida, F. chiloensis* ssp. *pacifica and F. chiloensis* ssp. *sandwicensis* are some important subspecies of *F. chiloensis*.

The *Fragaria virginiana* is another octoploid species of *Fragaria,* found throughout the central and eastern North America. The plants are slender and tall, producing runners profusely. The plants are mostly dioecious but some may produce a large number of hermaphrodite flowers also. Fruit is soft, round, light red, aromatic, with deeply embedded seeds and white flesh. Its plant and fruit characters are highly variable. The *F. virginiana* ssp. *virginiana; F. virginiana* ssp. *glauca; F. virginiana* ssp. *platypetala and F. virginiana* ssp. *grayana* are important subspecies of *F. virginiana*.

The *Fragaria iturupensis* is another octoploid species of strawberry, found in Kuril Islands of Japan. It has obovate, sub-glaucous leaves, similar in colour to those of *F. iinumae*. Its flowers are hermaphrodite and fruits almost spherical and red in colour. The *F. x ananassa* Duchesne, the common cultivated strawberry, is a hybrid of *F. virginiana x F. chiloensis*, and is also an octoploid.

There are also natural populations of higher-order polyploids found in coastal California. Hybridization between *F. vesca* and *F. chiloensis* has resulted in persistent 5x, 6x, and 9x colonies, allowing for introgression between the diploid and octoploid species. This group of hybrids is now recognized as *F. x bringhurstii*.

Chapter 5

Important Cultivars

Strawberry is adapted to different climatic conditions. It can be grown at high altitudes under tropical climate, in cold deserts and even under sub-tropical climate. Strawberry breeders have played leading role in the development of modern cultivars and standardization of various agro-techniques, which have resulted in the tremendous increase in strawberry production during the recent years. Several cultivars developed in different countries have been evaluated under different conditions. Nowadays, day-neutral strawberries are becoming favourite among growers because these are not affected much by length of the day and can be grown throughout the year. An ideal variety of strawberry should grow well in all types of climatic conditions, producing a large number of runners. It should produce higher yields of better quality fruits. Its fruits should be firm and of good dessert quality and colour. Besides, fruits should possess good processing quality, with ease in calyx removal. The berries should have better shape, tender texture and good flavour. An ideal ripening time and resistance to major insect-pests and diseases are other added advantages for an ideal strawberry cultivar. Keeping these objectives in mind, strawberry breeders have developed a large number of varieties throughout the world. The main characteristics of some of the strawberry cultivars are given hereunder:

Blomidon : Blomidon was developed in Canada, as a result of cross between K72-4 and Micmac. Its plants are vigorous, fruits large to very large, firm, glossy, medium to deep red exterior and medium-red interior, with white flesh. Calyx is moderately difficult to remove. It has tart flavour. Fruit ripens mid to late in season. It is moderately resistant to powdery mildew, Verticillium wilt, and resistant to Botrytis fruit-rot and red stele.

Cambridge Favourite : It is a cross between Avant Tout and Blackmore. An early to mid season variety, its plants are fairly large, moderately open and vigorous. The fruits are large, rounded and plump; pinkish-scarlet, turning light red on maturity. The flesh is firm, moderately juicy, pale-pink to pale-red with moderate flavour. Its fruits are suitable for processing purpose. It is resistant to mildew, but susceptible to Botrytis fruit-rot and many strains of red core. It produces runners profusely (Plate 2).

Cambridge Prizewinner : A cross between Early Cambridge and Howard 17, Cambridge Prizewinner is an early maturing variety bearing attractive red fruits. It is, however, not a heavy cropper. Plants are medium-sized, growing compact. Fruits are conical, occasionally wedge-shaped and bright red in colour. The flesh is moderately firm, pale-red and sub-acidic and has moderate flavour. Plants are susceptible to Verticillium wilt and red core but moderately resistant to Botrytis fruit-rot. It is a shy-runner producing cultivar.

Cambridge Rival : A cross between Dorsette and Early Cambridge, the plants of Cambridge Rival are tall, upright, fairly open, moderately vigorous with characteristic light yellow-green foliage. Fruits are large and often wedged, conical shiny, bright crimson, turning dark red when fully ripe but the tip of the fruit doesn't ripen. The flesh is firm, red and sweet with distinct flavour. It is susceptible to Verticillium wilt but resistant to most strains of the red core. The consumers do not like the intense dark colour of fruits and hence its marketing is a problem.

Cambridge Vigour : It is also a seedling selection of the cross between Dorsette and Early Cambridge. Plants are large to very large, upright and spreading. Fruits are fairly large but inclined to be very small towards the end of the picking season. Fruits are sharply conical in shape, look very attractive with orange-red to light red colour, turning scarlet when fully ripe. Flesh is firm, juicy, moderately sweet and with distinct flavour. It is resistant to some strains of red core, but very susceptible to Verticillium wilt and mildew and is highly

Plate 1. Strawberry plant

Plate 2. Cambridge Favourite strawberry

sensitive to drought. It produces plenty of runners and may grow too vigorously on good soils.

Canoga : It was released from the New York State Agricultural Experiment Station in 1979. The plants are vigorous and upright with dark bluish-green foliage. The fruits are exceptionally large, oblate in shape, and characterized by an intense, brick-red colour. The flesh is dark red throughout and is very firm. Canoga is susceptible to red stele and Verticillium wilt.

Catskill : It was developed at Geneva, from a cross between Marshall and Premier. The plants are vigorous having bright green foliage. The large attractive red fruits are good for the local markets. The fruits are very soft and have poor shipping quality. The berries darken on holding and are rated unsuitable for processing purpose. The plants are fairly resistant to Verticillium wilt, powdery mildew, leaf scorch and leaf spot diseases.

Chandler : It is highly acceptable present day cultivar of northern plains of India. It was developed in California by crossing Douglas and Cal 72.361-105. The fruits have exceptionally high dessert quality with outstanding colour, flavour and texture. The fruits are large (15-20 g), have low TSS (7-8%) and are quite resistant to physical damage from rain. This is a high-yielding, short day variety (Plate 3) and does exceedingly well under tropical and subtropical climates of India.

Dover : A cross between Florida Belle and Fla. Sel. 71-189, Dover is a high-yielding variety, bearing conic and slightly flattened fruits. They are generally smooth and uniform in shape with slightly recessed seeds. Ripe fruits develop attractive deep red colour. The fruits have a strong epidermis and firm flesh, which gives them excellent resistance to bruising during harvesting and shipping. Dover is highly resistant to anthracnose. It is good for fresh markets.

Earlidawn : A cross between Midland and Tennesse Shipper, Earlidawn produces fruits of good size and bright red colour. The

Plate 3. Fruits of Chandler strawberry

Plate 4. Fern strawberry

fruits are firmer than Catskill or Premier, the commercial cultivars of USA. The berries retain their good colour, texture and flavour when frozen. Earlidawn is a productive and a very early variety. It is susceptible to powdery mildew and highly susceptible to Verticillium wilt but resistant to fruit-rot.

Elista : The plants of Elista are moderately vigorous, upright and little streading. The fruits are large to medium, round conical to conical, sometimes wedge-shaped, orange red becoming scarlet with red flesh at maturity. The berries have good flavour but are little acidic. It is a heavy cropper, though the fruits of later crops are smaller in size. Fruits do not extend beyond leaf canopy but are well exposed. It shows some resistance to Botrytis rot and powdery mildew diseases. It may be of interest for processing purpose because of its deep coloured fruits.

Etna : Etna, a mid season variety, is a cross between Marlate and Belrubi. The plants are vigorous and spreading. The fruits are large and develop light red colour on maturity. It can be grown under glasshouse conditions as well as in open fields. It is a high yielding, medium to late ripening variety with resistance to Verticillium wilt.

Fern : It is an early day-neutral cultivar developed in California. Its plants are vigorous but grow erect. It bears large, conical-shaped fruits. The flesh is firm with excellent flavour. It is suitable both for fresh market and processing purpose (Plate 4).

Gilbert : The plants of Gilbert are vigorous, producing an abundance of runners which root readily. It bears very large, firm and red fruits. The cultivar is winter-hardy and resistant to leaf scorch. In addition to fresh markets, its fruits are also suitable for freezing.

Gorella : A cross between Jespa and US-3763, Gorella plants are moderate in growth, sparse with coarse thick foliage. The fruits are very large, conical, crimson red, sometimes with a green tip, and well exposed beneath the upright foliage. Flesh is bright crimson, juicy with moderate flavour.

Grenadier : A progeny of a cross between Valentine and Fairfax, Grenadier produces fruits of the same size as of Premier or larger than it. The very firm, dark-red berries are not suitable for all markets but are very good for table purpose. The berries are also good for processing as frozen whole or sliced berries and as jam. It is a mid-season variety and yields well. It is fairly susceptible to powdery mildew.

Honeoye : A progeny of Vibrant and Holiday, it produces vigorous plants having upright growth habit. It produces runners abundantly and is very winter-hardy. The fruits are large and attractive, uniformly conical in shape and have a bright red glossy appearance. The fruits are moderately firm with relatively tender skin (Plate 5).

Jewel : It was selected in 1971 from the progeny of NY 1221 x Holiday in USA. It is a hardy variety and is consistent in cropping with fruit yield similar to Guardian. The berries are large (14-15 g), bright red in colour and moderately firm, possessing high quality both for fresh and frozen uses. The berries have very high and attractive flavour. Jewel is resistant to post-harvest fruit rot.

Latestar : A cross between Lateglow and Allstar, Latestar has been released by US Department of Agriculture for commercial cultivation. Latestar was named after its parents Lateglow and Allstar. Its plants are vigorous, having very good runnering capacity. The fruits are attractive, large with firm flesh and tough skin. The colour of fruits is glossy red. The flesh is light red; pleasant and mild, slightly acidic with fresh flavour. The plants of Latestar are resistant to most of the common leaf diseases and fruit rots but susceptible to leaf blight. It grows well either on light or heavy soils. It is a late season cultivar and good for local markets.

Lester : It is a seedling selected from the Raritan X MSUS-3413 cross in USA. Its plants are medium to small, runnering moderately to form a plant bed of medium density. The leaflets are medium dark green on the upper surface and light waxy green on the lower side. Primary fruits are large (18-20 g) and are generally wedge to conic in shape. However, secondary and late ripening berries are medium in size and short conic in shape. However, fruit shape is very symmetrical in outline. Skin colour of fruits is glossy deep red; flesh colour varies from medium red in the outer surface to pink or whitish at the core. Fruit skin and flesh are firm. Fresh fruit flavour is mild, sweet and pleasant. Fruits are highly tolerant to fruit-rot.

Mars : Mars is a descendant of noteworthy cultivars like, Howard 17, Beaver, Dorsett, Rockhill, Dunlap and Sparkle. It has been named 'Mars' after the bright red planet. Mars is a vigorous and productive cultivar, producing dark green crown of uniform green colour. It produces a large number of runners. The berries are medium-sized, ripen in mid season and are oval shaped bright red with a dark green calyx. It is resistant to all foliage diseases like leaf spot, leaf scorch etc. The fruits are good for freezing. Since Mars has quite concentrated ripening, it is suitable for mechanical harvesting.

Nanaimo : Nanaimo is a cross between Totem and Korona. Plants of Nanaimo are vigorous and have high runnering capacity. It has a high level of winter hardiness. Leaves of Nanaimo are medium in size, light green in colour with symmetric serrations on both halves of the leaflets. The fruits are medium-sized and bright medium red both externally and internally. The fruits are mostly wedge-shaped and have prominent constriction near the tip. It is slightly susceptible to powdery mildew, leaf scorch, red stele root-rot and pre-harvest fruit-rot. It is particularly useful as a sliced product for dessert packs that have excellent appearance and flavour but not good for individual quick-freeze because the fruit does not have the firmness to maintain its integrity.

Pajaro : Pajaro, a seedling of Cal 63.7-101 X Sequoia, was released during 1979 in California (USA). The fruits are of high dessert quality, large, symmetrical, attractive red and firm. The fruits are, however, quite susceptible to physical damage from rains. It is an early variety, giving high yield of better quality fruits. It is resistant to fruit-rot, wilt and leaf scorch (Plate 6).

Plate 5. Developing fruits of Honoeye strawberry

Plate 6. Fruits of Pajaro strawberry

Paros : A cross between Marmolada and Irvine, Paros was released for commercial cultivation in 1998. It produces conical, elongated to conical-spherical, highly speckled, red orange fruits, which are mid-early ripening, weighing 28-31 g each. On an average, a plant yields 820 g fruits in a season.

Pelican : It has been released only recently for commerical cultivation in Florida (USA). Its plants are large, vigorous and erect. Leaves are large, cupping upward at the margins, medium green on the top and light green beneath. Leaflets are broad, ovate with round and tipped serrations. Fruits are large, medium firm, very long, wedge-shaped with a glossy red exterior. Interior flesh is pink and melting (uniform and smooth), juicy with balanced sweet and acidic flavour. The skin may bruise or crack under some conditions. It responds well to raised bed plasticulture system. It has outyielded Chandler and Sweet Charlie at many locations. Its plants are highly resistant to anthracnose and red stele root rot diseases.

Premier : It was introduced by A.B. Howard to replace Ontaria strain, and continued to be one of the leading commercial varieties in Ontario. It is a cross between Crescent and Howard 17. The Ontaria usually produces soft fruits, which darken during storage and are thus not considered good for processing purpose. Moreover, virus-free stocks of Ontario strain are not usually available. The plants of Premier are fairly resistant to Verticillium wilt, powdery mildew, leaf scorch and leaf spot diseases. Moreover, its fruits are firm and do not darken during storage and are thus considered good for processing purpose.

Primetime : A cross between MDUS 4377 and Earliglow, Primetime was released for commercial cultivation in USA during 1996. The name Primetime has been given because it ripens during the "prime" season in USA. Its plants are vigorous and produce profuse runners. It bears attractive, large, symmetrical and good quality fruits. The fruits are medium in skin firmness and toughness, with bright scarlet skin and pink flesh having pleasant, mild and mild aromatic flavour.

It is a June-bearing mid-season variety. Primetime is resistant to red-stele root rot, several leaf and fruit-rot diseases. It has been recommend for local markets but not for processing purposes.

Puget Reliance : Puget Reliance was selected from a cross between WSU 1945 and BC 77-2-72 in 1983. Its plants are vigorous with an erect growth habit. The leaves are medium in size and the petiole hairs are irregularly perpendicular to the axis of petiole. The fruits are usually smooth, glossy and symmetrically conic with inserted seeds. The calyx can be easily removed at picking. It is susceptible to aphid but highly tolerant to viral diseases, leaf scorch and anthracnose fruit-rot. Similarly, powdery mildew and red stele are not problems in Puget Reliance.

Redcoat : A seedling selection of a cross between Sparkle and Valentine, Redcoat is a high-yielding variety, introduced in Canada during 1957. Redcoat retains good berry size throughout the season. It is one of the most outstanding varieties for fresh market. The fruits are bright red, very firm and attractive, and maintain their appearance in the storage. The plants are, however, susceptible to Verticillium wilt and leaf spot but resistant to powdery mildew.

Red Rich : It is also known as Red Glow and Hagerstorm's everbearing strawberry. It is a seedling selection from a cross between Market Surprize and Senga Sengana. It is a mid-early variety, bearing heart-shaped fruits of uniform shape and size. The fruits develop glossy bright red colour at maturity. This is one of the most promising everbearing variety of strawberry. The fruits are good for dessert and freeze well. It develops only a few runners. The flesh is firm, light red and has acid-sweet flavour. Yield is good with a large proportion of top quality fruits suitable for fresh consumption, freezing or making jam.

Redcrest : A selection made from a cross between Linn and Totem, Redcrest plants are vigorous and spreading. The plants produce an adequate number of runners. Its foliage is slightly lighter green in colour. Most of the flowers are produced at or above the canopy, but frost has not been a serious problem because it is a late ripening variety. The fruits are medium-sized (10-12 g), having attractive red colour. Plants are resistant to blight and red stele root-rot. The yields are similar or slightly less than Totem (one of its parent), the most popular processing cultivar of USA. Redcrest has been primarily selected for high quality processing purpose as it has attractive colour and good texture to provide excellent sliced or individually quick freeze (IQF) packs and preserves.

Redgauntlet : A progeny of New Jersey 1051 and Auch. Climax, Redgauntlet is a high yielding variety. It produces good-sized berries throughout the harvesting season. The plants are tall and vigorous but their growth is sometimes weak, particularly in poor soil conditions. Fruits are large to very large, rounded to spherical-wedged, broad, blunt, often uneven, with few seeds having attractive scarlet and dark crimson colour on maturity. Flesh is firm, fairly juicy, and deep scarlet with fair flavour. The fruits are well exposed but the fruit stalk is sometimes tough. It shows field resistance to some strains of red core and to Verticillium wilt. This is less prolific in runner production than the Favourite. It responds well to warm conditions and sheltered sites.

Scott : Scott is a cross between Sunrise and Tioga. It has been named after D.H. Scott, an internationally famous strawberry breeder. It is a vigorous and a prolific runner maker. The berries are large, ripen in late mid-season, with firm flesh and skin that resists bruising. The fruit is attractive and wedge-shaped. The fruit has bright red external and solid light internal colour. The flesh is firm with mild and pleasant aroma. It is resistant to *Phytophthora fragariae* and partially resistant to Verticillium wilt, leaf scorch and powdery mildew but susceptible to anthracnose.

Selva : It is a day-neutral variety, which produces fruits both in open, under low plastic cloches or in peat bags under greenhouses. The yield of Selva is less than Rapella, but average fruit weight is more than Rapella. This variety is doing well in sub-tropical zones of India (Plate 7).

Senga Sengana : It is a seedling selection from a cross between Markee and Sieger. It was introduced in Europe in 1952 and in India in 1980. It produces healthy and vigorous plants. Its runners begin to form during August. The fruits are attractive red with firm flesh and good freezing quality. It does well under hilly zones of India (Plate 8).

Sparkle : Sparkle is also known as Paymaster and has been selected from a cross between Fairfax and Aberdeen. Its plants are vigorous and spreading. The medium-sized, firm, rather dark and attractive berries are ideal for dessert purpose and freezing. It is a fairly high-yielding variety. The plants are moderately susceptible to Verticillium wilt, leaf scorch and leaf spot but resistant to powdery mildew (Plate 9).

Surecrop : Surecrop is a result of a cross between Fairland and Maryland. Its plants are vigorous and spreading with dark green foliage. The fruits are medium-sized, conical in shape, retaining their uniform size throughout the season and storage. The fruits are attractive red, firm, good for desert and satisfactory for freezing. The plants are resistant to certain races of red stele and fairly resistant to Verticillium wilt, leaf scorch and leaf spot diseases.

Sweet Charlie : Sweet Charlie is a cross between FL 80-456 and Pajaro, which was released for commercial cultivation in Florida (USA) during 1997. Sweet Charlie has been named in honour of the Late Charles M. (Charlie) Howard, a famous plant pathologist at the University of Florida. The plants of Sweet Charlie are compact and smaller with good runnering capacity. The leaflets are generally slightly cupped, medium to dark green, semi-glossy, scabrous and obovate. Primary fruits are usually wedge-shaped, whereas, fruits of late flush are conical to wedge-shaped. Fruits are large (average weight 17 g). External fruit colour is orange-red and flesh colour is orange streaked with white. The achenes are greenish-yellow and slightly recessed. The fruits are moderately firm and have a high ascorbic acid content (53 mg/100 g fruit weight) and medium to low

Plate 7. Attractive fruits of Selva strawberry

Plate 8. Senga Sengana strawberries

TSS (7.0 per cent) and low titerable acidity (0.66 per cent). Fruits have distinctly sweet flavour. It is a short-day and early fruiting cultivar and grows well in relatively mild winters. It is resistant to anthracnose fruit-rot but susceptible to Phompsis blight, powdery mildew and to two-spotted spider mite (*Tetranychus urticae*).

Tillikum : Tillikum is a day-neutral (everbearing) cultivar, selected from a cross between Totem and Sel. 65.65-601. Its plants are vigorous and medium in size, bearing numerous inflorescences per plant. Each inflorescence generally has 15-20 flowers. Fruit size varies from small to medium (6-9 g). Skin colour is glossy and deep red at maturity. The flesh is deep red throughout. Flavour is acidic and good when eaten fresh. Foliage of Tillikum is susceptible to leaf spot but moderately resistant to powdery mildew. Roots are moderately resistant to red stele. It produces good runners if plants are set early and blossom clusters are removed.

Tioga : A cross between Fresno and Torrey, Tioga has short harvest season. Its fruit size decreases with the advancement of season. It is also somewhat difficult to pick for the fresh market because the fruit tends to 'cap' easily and come off without calyx. It is suitable for processing but not for fresh market. However, its plants grow well in the mid hills of Himachal Pradesh. It is ready for harvesting in second week of April in mid hills of Himachal Pradesh.

Torrey : A sister seedling of Tioga, the plants of Torrey are of medium vigour, producing red fruits. The fruits are large-sized with firm flesh and cap easily than Tioga. Under mid hills of Himachal Pradesh, it ripens by first week of April, a little earlier than Tioga. The fruits are good for fresh market and processing as well.

Tribute : It is a seedling selection of the cross EB 18 x MDUS 4258. This is perhaps the first day-neutral (everbearing) cultivar developed in USA. The plants are medium to vigorous. The fruits are large to medium with shape irregular to symmetrical with short conic wedge and have pronounced shoulders. The fruit colour is glossy bright red, not quite as deep as that of Tristar. Flesh is solid, medium red in colour. It has slightly acidic taste when eaten fresh. Flesh and skin texture is quite firm. It gives very high yield in matted row system of planting. Roots are highly resistant to red stele rot and tolerant to Verticillium wilt, while leaves are resistant to powdery mildew, blight and leaf scorch diseases.

Tristar : It is a sister seedling of the Tribute. It is also a day-neutral cultivar. The plants are medium to small with medium vigour. The plants bear heavy, very early spring crop of small to medium-sized, symmetrical, short conic berries with a marked shoulders and reflexed cap at maturity. Fruit flesh and skin are firm. Skin colour is a glossy red with a "star" like reflecting pigment around the achenes. Flesh colour is deep red and the flavour is excellent when eaten fresh. It is resistant to red stele root-rot with a high degree of tolerance to Verticillium wilt.

Vantage : A cross between Tioga and Vestar, Vantage is a high yielding cultivar, the fruits of which are suitable both for fresh and processing purposes. The fruits are bright in colour, firm with good flavour. The plants are resistant to *Botrytis cinerea* and highly resistant to Verticillium wilt.

Vesper : This is a late ripening American cultivar. Its fruits are large, conical and elongated with deep red colour. The average fruit weight is 9-11 g. The plants are vigorous and are not readily depressed by viral diseases.

In India, many cultivars have been introduced from USA, the then USSR, Italy, Australia and Spain by NBPGR, Regional Station, Phagli, Shimla (HP), Department of Horticulture, Government of Himachal Pradesh, Department of Horticulture, Government of Jammu and Kashmir, Department of Agriculture, Government of Uttar Pradesh, VPKAS, Almora (Uttaranchal). Several investigators have studied the performance of different strawberry cultivars in different locations in India. Of them, Tioga, Torrey, Douglas, Fern, Chandler, Senga Sengana, Pajaro, Redcoat, Tribute and Tristar have been found suitable for cultivation in different regions of India.

Chapter 6
Soil and Climate

For growing strawberries profitably, ideal soil and climatic conditions are important. Before starting the business, selection of proper site, soil, availability of water and other requirements should be ensured.

Location and site : Selecting a good site with a suitable soil is the first important step in the successful growing of strawberries. The factors most important in selecting a site for commercial strawberry plantations are (a) accessibility of markets, (b) transportation facilities, (c) adequate labour availability (d) community interest and (e) appropriate climate. If it is to be grown for general market, it is usually best to select a site among other growers. Before starting strawberry cultivation on a large scale, the grower should have reasonable assurance that there will be adequate supply of labour, especially during harvesting season. Other important factors in selecting the site are water, proper slope and better soil type. Gentle slopes are best for strawberry cultivation because they ensure better drainage, whereas on steep slopes, the soil erosion can be a serious problem, rendering its cultivation less attractive. However, if steep slopes are used, contour planting may be followed with or without terraces in order to conserve soil and moisture, though contour planting is considered to be a good practice on sites with only a slight slope.

Soil

Strawberry can be grown on a variety of soils, ranging from heavy clay to gravel soils, but it prefers light porous soils, which are rich in humus, because its roots are confined to the top 15-20 cm layers of soil. In light soils, frequent irrigation is required for proper establishment of the runners and to maintain good berry size and fruit quality. In heavy soils, development and penetration of roots of the runners is inhibited. Though strawberry is not sensitive to soil reaction, it prefers slight acidic soils with a pH of 5.8-6.5. On alkaline soils, strawberry may not establish properly and it may result in leaf chlorosis due to lack of iron absorption by the roots. Similarly, strawberry suffers badly in poorly- drained soils. Waterlogging, even for a limited period, may kill the whole plantation. Poorly drained soils also encourage fruit-rot diseases, affecting plant growth, and fruit yield and fruit quality adversely. Thus, it is essential to improve the surface drainage before planting, otherwise it may result in complete failure of the crop.

Climate

The cultivated strawberry is a hybrid of two highly variable octoploid species. Therefore, it is possible to raise it profitably under extremely different climatic conditions. However, among different climatic factors, temperature and day length have considerable effect on growth and yield of strawberry, probably through the control of plant hormones. The different climatic factors are:

a. *Light* : Flowering in strawberry is strongly influenced by photoperiod, temperature and photoperiod x temperature interaction. On the basis of photoperiodic requirements, strawberry cultivars have been categorized as short-day (SD), long-day (LD) or day-neutrals (DN). In short-day cultivars, floral induction occurs with photoperiods of less than 14 hours, though these cultivars flower continuously regardless of day length but the environmental temperature should be less than 16°C. The day-neutral cultivars, oftenly called as **everbearer**, flower continuously, regardless of day length. Although temperature also modifies the photoperiodic response but day-neutral cultivars are less sensitive to high temperature than the short-day cultivars. The third group is between short-day and day-neutral group, in which flowering takes place under long day conditions. Under northern Indian conditions, fruit-bud differentiation in strawberry starts by mid-November, though it is genotypic dependent. Variation in light quality can affect floral inductive conditions and also delay or inhibit leaf petiole length and stolon formation in some cultivars.

Generally, flowering in strawberry occurs under the following photoperiodic regimes, short light period (10 hr) with long dark period (14 hr), short light and dark periods (10 hr), and long light and dark periods (14 hr). Thus, the duration of dark period rather than the light period is the factor, which controls the floral initiation in the strawberry. Photoperiod has a marked effect on strawberry vegetative growth, morphology and yield. It has been repeatedly observed that stolon formation, petiole length, leaf area and yield increase with the increasing photoperiod, though the effects are cultivar-specific.

Present-day strawberry (*Fragaria x ananassa*) is grown extensively in temperate and to some extent in subtropical culture, i.e. short-days in autumn and hard winter, or to meridional culture, i.e. long days in autumn and moderate winter. The cultivars like Gorella, Redgauntlet and Senga Sengana grow under septentrional conditions, while Tioga and Torry are important cultivars of meridional climate. However, with the development of several day-neutral cultivars like, Selva, Fern, Hecker, Aptos, Tribute, Tristar, etc., it is possible to grow strawberry throughout the year, but only in temperate climate.

b. *Temperature* : The growth and development of strawberry are highly sensitive to variations in the air and soil temperature. Simple effects of temperature as well as interactions often are cultivar and species dependent. An optimum growing temperature of 15° C has been reported for most of the strawberry cultivars and species, though it grows well at a temperature range between 20° C and 26° C. However, low temperature injury to the crown increases with the increase in the duration of low temperature.

Spring or autumn frost often results in injury to strawberry reproductive organs. Cold injury to open flowers often occurs at 2° C, although flowers of several cultivars survive even at -4.5 °C. Susceptibility to cold may vary with cultivar and the degree of ice nucleation that occurs in the tissue.

c. *Photoperiod and temperature interaction* : The photoperiod and temperature interaction play a vital role in the regulation of vegetative and reproductive phases of strawberry. Generally, short-day conditions favour floral initiation and inhibit stolon formation, regardless of the temperature. However, optimum temperature for stolon formation and development of floral initiation is photoperiod dependent. The fruiting season for short-day cultivars tend to be longer in areas with mild winter due to longer growing season in which photoperiods are less than 14 hours and chilling injuries are also few. In general, critical photoperiod for inhibition of flowering and that for promotion of vegetative growth, such as increased petiole length, leaf size, and stolon formation are different.

d. *Salinity* : Strawberry is one of the most sensitive crops to salinity. In areas with low rainfall and marginal water quality, excessive salt accumulation may occur in the soil. High salt concentrations inhibit new root initiation and root growth, resulting in poor anchorage of the plant, loss of feeder roots, leaf necrosis, and stunted growth of plant or even in plant death but reduction in fruit yield often occurs before visual symptoms of salt damage are apparent. Higher concentration of chloride is especially damaging and the sodium is comparatively less toxic. Proper irrigation is important for minimising salt damage to strawberry. Furrow irrigation in localities affected with salts may be avoided and sprinkler or drip irrigation should be encouraged. Sensitivity of strawberry to salinity can vary, depending on environmental conditions. For example, detrimental effects of salinity are often more pronounced under hot, dry conditions than under cool conditions.

e. *Waterlogging* : Strawberry is highly sensitive to excessive soil moisture and poor soil aeration. Moreover, strawberry roots are susceptible to a number of pathogens that survive in poorly drained or waterlogged soils. The feeder roots of strawberry are highly sensitive to poor soil aeration and die rapidly when exposed even to short periods of flooding. Thus, the sites selected for strawberry cultivation should have proper drainage system.

Altitude : Strawberry can be grown even up to an elevation of 1,500 feet above sea level. However, plant development is delayed at such elevations as compared to low altitude. Thus, low lying valley areas are most suited for its cultivation.

Acidic fog : Strawberry plant appear to be relatively tolerant to acid fog. However, sulphuric acid rain may result in necrosis of leaves, petioles, flowers and fruits, without affecting the fruit number.

CO_2 concentration : The concentration of CO_2 plays a vital role in strawberry production under protected cultivation. Exposure of the plants to elevated CO_2 concentration typically enhances vegetative growth and yield, although concentration above 1000 (μM) may result in stomatal closure and decreased vegetative growth. In general, CO_2 concentration of 750 (μM) is optimum for most of strawberry cultivars grown under greenhouse conditions.

Czech Republic	17
Finland	10
France	[illegible]
Germany	[illegible]

Chapter 7
Area and Production

Most of the European countries have been growing strawberries commercially since the eighteenth century. Nowadays, about 60 per cent of the world's production is only in Europe. In Canada, main strawberry growing regions are Ontario, British Columbia and Quebec. Its commercial cultivation in USA was started at the beginning of eighteenth century and within twentyfive years, it gained significant momentum. Strawberry is also grown in Israel, Japan, Turkey, Australia and New Zealand (Table 3). Poland produces only processing strawberries. In India, NBPGR, Regional Research Station, Phagli (Shimla) and IARI Regional Horticultural Research Station, Shimla (Himachal Pradesh) during sixtes, first introduced strawberry. But the early efforts made to popularize its cultivation in H.P. and U.P received a setback on account of poor adaptability of the introduced cultivars, low returns per unit area and lack of the technical know-how on different cultivation aspects of strawberry. However, after the introduction of Tioga, Torrey, Elista, Chandler, Shasta, Douglas, Fairfax, Senga Sengana etc. cultivars and standardization of different agro-techniques, its cultivation got a boost during the last decade. Now its area and productivity are increasing day by day. Presently, it is being grown in Shimla, Solan, Bilaspur, Kangra, Kullu, Palampur (HP); Dehradun (Uttaranchal), Saharanpur, Muzaffarnagar, Ghaziabad (UP); Hoshiarpur, Ludhiana; Jalandhar, Patiala (Punjab); Gurgaon, Hisar, Karnal (Haryana); Bangalore, Coorg (Karnataka); Kodaikonal, Ooty (Tamil Nadu); Pune, Mahabaleshwar (Maharashtra) on a small scale.

The total world production of strawberry during 1998 was 2.6 million tonnes (Table 3). Europe produces about 1/3rd of the total

Table 3 : World scenario of strawberry production during 1998 ('000 MT)

Continent /Country	Production
World	2,601
Africa	55
Egypt	40
Morocco	10
South Africa	4
N C America	870
Canada	27
Mexico	100
USA	742
South America	63
Argentina	8
Chile	16
Colombia	15
Peru	14
Venezuela	4
Asia	516
China	7
Iran	13
Israel	14
Japan	200
Korea Republic	151
Lebanon	13
Turkey	110
Europe	1,081
Austria	16
Bel-Lux	36
Bulgaria	5
Czech Republic	17
Finland	10
France	78
Germany	82

Continent /Country	Production
Greece	7
Hungary	13
Italy	81
Netherlands	25
Norway	7
Poland	150
Romania	12
Russian Fedral Republic	115
Spain	313
Switzerland	9
UK	31
Ukraine	26
Yugoslavia	19
Oceania	16
Australia	12
New Zealand	5
USSR	98

strawberries of the world. Among different countries, Spain, Poland, Germany and France are the major strawberry producers of the world. USA, Mexico, Egypt, Japan, Italy, and Russian Federal Republic also produce sizable amount of strawberries (Table 3). Unfortunately, the data on its acreage and production in India is not available. However, due to pressing demand of farmers and consumers, adoption of modern and standardized agro-techniques, introduction of day-neutral varieties and use of protected cultivation, both the area and production in India have increased substantially during the past few years.

Chapter 8
Plant Propagation

Strawberry can be propagated through sexual (seed) and asexual (vegetative) means. Propagation by seed is not considered as a viable method of propagation for cultivated strawberry as the plants raised through seeds do not come true-to-type. However, strawberry is usually propagated through runners. Nowadays, its large-scale propagation by tissue culture is being used widely.

Seed propagation : Like many other fruit crops, propagation of strawberry through seeds is not considered as a right method of propagation. However, hybrids are first raised through seeds only. A single berry produces thousands of seeds, so a strawberry breeder has an advantage over other fruit crops in the sense that a number of hybrid seedlings can be produced with a single attempt. Seeds are very small and do not germinate properly unless they are stratified for a certain period at a given temperature. The effective chilling temperature (the temperature required for breaking rest) ranges from -2° C to 6.5° C but a chilling temperature of 9.5° C to 10° C is apparently most effective. The duration of chilling period is usually 2-4 weeks. Further, the treatment of the seeds with H_2SO_4 or GA_3 and thiourea also breaks the seed dormancy and facilitates better seed germination.

Vegetative propagation : The stolon, a creeping stalk, is produced in the leaf axil and grows out from the parent plant during summer. At the second node, a runner plant is formed and a new stolon arises on the runner plant to continue the runner train, which often branches (Plate 10 and 11). Initially, the runner plant produces a few roots and thereafter makes excessive fibrous roots (Figure 2). Thus, the strawberry naturally propagates itself by vegetative method of runner production. After the runner plant has acquired sufficient growth and roots, it may be separated from the mother plant and can be planted elsewhere. Although, propagation by runners perpetuates all the characters of the mother plant, viral diseases are quite often transmitted through runners. Thus, for runner production, a separate bed should be used. The site and soil where strawberry had not been grown for the last 3-4 years should be selected for the runner production. Soil should be tested for viruses and other soil-borne pathogens. Virus-free or tested clonal material of the required variety should be used. The planting for runner production should be done at 1.2 m x 1.2 m or at 1.8 m x 1.8 m distance. All other cultural practices like irrigation, nutrition, weeding and mulching should be strictly followed till the runners are lifted from the bed. Usually, 30-40 runners from a mother plant is considered satisfactory rate of production. Runner production is not a problem in June-bearing varieties. However, in everbearing strawberries, runner production is very low because the flower bud development continues throughout the growing season. This creates a problem to the plant propagators or the nurserymen not only because the runner production is poor but also because the flower buds are to be removed by hand throughout the growing season to prevent fruiting and to favour runner production. In India, strawberries produce runners frequently in hilly regions. However, in plains, runners are not produced frequently and if at all produced, it is difficult to protect them from high temperature in May-June.

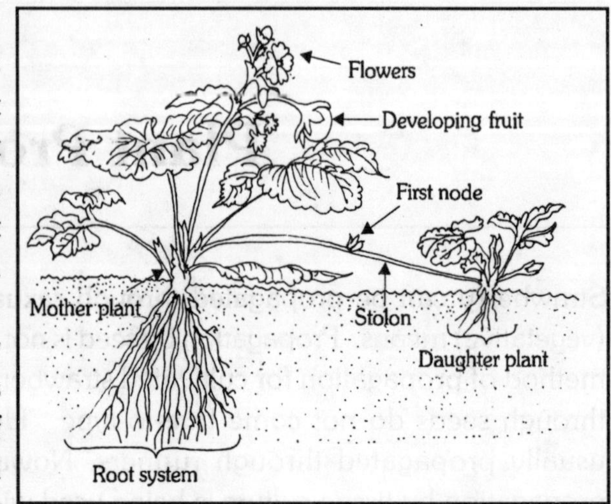

Figure 2. Runner production in strawberry

Plate 9. Shinning fruits of Sparkle strawberry

Plate 10. Runner initiation in strawberry

Runner production is invariably a varietal character, however, it can be enhanced with the treatment of different growth regulators. A single spray of GA_3 (50-100 ppm) to the new and one-year-old plantings of everbearing strawberry inhibits flowering, promotes runner development and increases the yield of marketable runners in the following years. In contrast, runner production is inhibited by soil application of paclobutrazol.

The runners are mostly lifted during September-October. Sometimes, the lifting continues upto March to meet the demand for spring planting. However, for spring planting, dormant runners are cold stored. For this purpose, runners are usually lifted in late November-December and stored until planting, i.e. June-July next year.

Storage of runners : Runners should be stored when they are fully developed and dormant. These should be shaken well to remove soil from roots and packed in bundles of 200-250 each in polyethylene bags. These bags may be kept in shelves in a cold storage room, maintained at a temperature of 0°C or so. The runners can be kept in good condition for three months at 0° C or -1° C. Dipping runners in 0.2 per cent thiram and or removing most of old leaves before storage is useful for enhancing the storage life of the runners. The cold stored runners are usually healthy, however, rotting of leaves and crown may take place. It usually happens due to lack of proper cooling and infection by fungi. The runners should be removed from cold storage during April-May and planted as soon as possible to avoid injury from the heat.

Micropropagation : Micropropagation refers to the propagation of the plants from the smallest plant part under controlled nutrient, environmental and aseptic cultural conditions. Strawberry is perhaps the first fruit crop in which micropropagation technique was first standardized. Nowadays, its large-scale commercial propagation is being done through tissue culture. Strawberry is the fruit plant, which requires very high number of plants/ha (50,000-60,000) and this demand can easily be met through micropropagation. Different explants like meristem tip, anthers and immature embryos and first axillary buds of stolon can be used to produce millions of plants in a year. The culture media like Knoop's solution, Murashige and Skoog's medium (MS) and Linemaier-Skoog (LS) media with agar, sucrose and hormones at desired levels have been standardized. The addition of growth regulators to culture medium has shown very stimulative effect on regeneration, growth and yield of tissue cultured plants. Usually, growth of tissue cultured plants in the field is very good, producing lust green foliage. Generally, these plants are free from diseases, producing high yield of quality fruits.

Advantages of micropropagation

1. In micropropagation, only very minute plant parts are used.
2. Its large-scale multiplication in limited time and space is possible.
3. It is a labour-intensive technique and it is possible to produce plants throughout the year.
4. The plants produced are healthy and usually free from diseases.
5. Everbearing strawberries even perform better under *in vitro* conditions compared to field propagation.
6. The tissue-cultured plants perform better in field compared to those propagated by conventional means.
7. The tissue-cultured plants have higher runnering capacity.
8. Screening of these plants against insect-pests and diseases is comparatively easy.

Chapter 9

Soil Preparation & Planting

Preparation of land : Before planting, the land should be cultivated through repeated ploughings and by removing weeds, stubbles etc. It is not advisable to follow commercial planting of strawberries on a land infested with perennial weeds, since it is difficult to eradicate them later. If perennial weeds are present, adequate cultivation in a fallow land may be done or weeds may be erradicated by applying suitable weedicides before planting. The soil should be cultivated for 1-2 years before planting strawberry to make the field free from white grubs and perennial weeds. Since most of the strawberry roots are confined to the top 30-40 cm of the soil surface, the soil should be made friable. Deep cultivation increases the capacity of plants to absorb more water and nutrients from the subsoil. Moreover, it is essential for the proper development of roots. The fumigation of soil is essential before planting since it makes soil free from soil borne pathogens. Fumigation with chemicals like methyl bromide (67%) or chloropicrin (33%) is useful before planting. Similarly, soil solarization is more effective than fumigation if followed scientifically.

The desirable doses of farmyard manure (FYM), phosphorus (P), $^1/_2$ N and potash (K) should be added during ploughing and the rest as top dressing. If strawberry is being grown on a same patch of land for the last many years, the humus content of soil should be increased before new planting. This can be done by applying 8-10 tonnes of manure to the crop preceeding strawberry. If manure is not available, two green manure crops preceding the strawberry may be grown and cultivated in the soil before its planting. Under Indian conditions, *dhaincha* may be grown as a green manure crop. Application of lime is very useful in strawberry cultivation. It not only supplies calcium and magnesium to the plants but also binds toxic aluminium. Its application is, however, essential in humid climate if the pH is below 5.0. For this, 1-2 tonnes of dolomite may be added in one-acre land to raise the pH from 5.5 to 6.5. However, its application should be avoided in arid climate. It is always better to lime the land one year before planting. After thorough ploughing, soil fumigation and addition of manures and fertilizers, the soil should be levelled and beds should be prepared in a way that water can run in either direction in furrows between the beds.

Planting

Planting can be done either on furrows or on raised beds in a particular system of planting. In India, raised beds are usually preferred. In some localities, planting is also done on furrows like potatoes. The planting may be done by hand but care should be taken to prevent damage and drying of roots of the runners. After removing runners from soil, these may be planted directly in field or placed in shadow before transplanting. It is better if runners are kept in a moist sphagnum grass. It prevents drying of roots and keeps runners in fresh condition for a longer period. Excessive wilting of runners may increase foliage loss and plant mortality, reducing plant size and affecting yield and fruit quality adversely. Planting should be done in early or late evening hours. It should, however, be avoided in sunny day and under wet or frost conditions. Immediately after planting, a light irrigation should be given.

Planting systems : Three planting systems are usually adopted for planting strawberries. They are hill, spaced row, and matted row system (Figure 3). In some localities, single hedge, double hedge row, single row or double row hill are also followed. Similarly, in backyard plantings, barrel or pyramids are also used (Figure 4). Some of the commonly followed systems are described hereunder :

1. ***Hill system*** : This system is commonly followed in cultivars developing few runners. In this system, the plants are grown either in single or double rows on the raised beds. The beds are usually made 15-20 cm high and runners are set 20-25 cm apart

in twin rows, 30-35 cm apart and a distance of 75-100 cm is kept between two rows. With this system, a small garden tractor or a field tractor can be used for tillage, which can reduce the excessive cost of hand labour. In the home gardens, where hand labour is used, the rows can be spaced closer or about 45 cm apart. Sometimes two rows are set 40-60 cm apart, then a wide space is left and another two rows are set. In some cases, triple rows are set. However, these are only modifications of hill system with the plants set at the same distance apart in the row in each case. This system demands a much higher economic input and a large number of runners are set per unit area and thus makes removing of runners difficult.

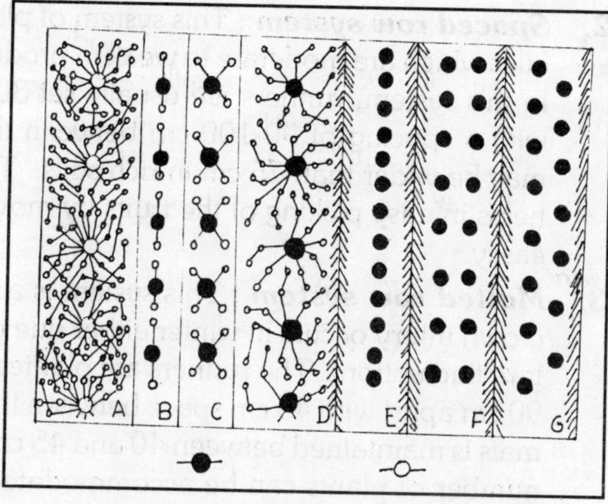

Figure 3. Systems of training strawberries

(A. Matted row, B. Single hedge row, C. Double hedge row, D. Spaced row, E. Single row hill, F. Double (opposite) row hill, G. Double (alternate) row hill)

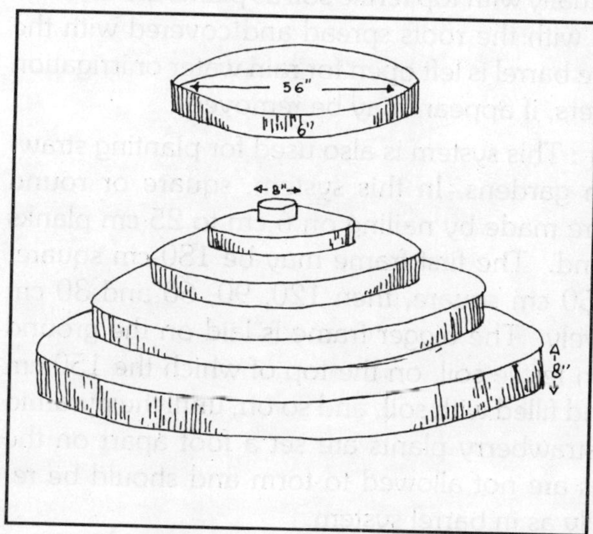

Figure 4 : Structures for growing strawberries in backyards

2. ***Spaced row system*** : This system of planting is used for varieties which are moderate to weak in producing daughter plants. In this system, runners are usually set 30-50 cm apart in rows with a spacing of 90-100 cm between the rows. These rows may be wider than 75 cm in rich soils. This system of planting helps in easy picking of the fruits by moving between the rows easily.
3. ***Matted row system*** : This system is adopted in areas where crown injury occurs in winter either due to freezing or temperature fluctuations. The runners are planted along the rows about 90 cm apart with 45 cm space between the plants. The width of mats is maintained between 40 and 45 cm. In this system, more number of plants can be accommodated per unit area, which may give higher yield under suitable conditions. However, overcrowding may cause higher fruit-rot and thereby a reduction in fruit yield. This system is suitable for obtaining heavy crop of medium sized fruits for processing purpose.
4. ***Barrel system*** : This system is used in backyard planting, where land is limited. A 25 or 50 gallon wooden barrel can be used for this purpose. Holes are bored in the side of the barrel. The barrel is filled gradually with top fertile soil as plants are inserted through the holes with the roots spread and covered with the soil. The top of the barrel is left open for rain water or irrigation as needed. Runners, if appear, may be removed.
5. ***Pyramid system*** : This system is also used for planting strawberries in kitchen gardens. In this system, square or round wooden frames are made by nailing on 6 cm to 25 cm planks together end to end. The first frame may be 180-cm square, the next about 150 cm square, then 120, 90, 60 and 30 cm square consecutively. The bigger frame is laid on the ground first and filled with fertile soil, on the top of which the 150 cm frame is placed and filled with soil, and so on, until the pyramid is formed. The strawberry plants are set a foot apart on the steps and runners are not allowed to form and should be removed periodically as in barrel system.

Planting density

The planting distance and planting density depend on several factors like, physical conditions of the soil, growing conditions, cultivar and planting system. A considerable amount of work has been done on these aspects and it has been found that strawberry yield is positively related to planting density. However, the increasing density has adverse effect on number of flowers per shoot, fruit yield and quality. Due to higher planting density, there may be more competition for nutrients, water and light, which play a vital role in flower initiation. The branching of crown is also inhibited under dense planting, which is basically an important factor in determining yield potential of strawberry. Cultivars producing single crown can be planted in dense planting but reducing planting distance further also affects leaf size, number and crown. The plants raised on lower density (25 plants/ m^2) usually produce more number of flowers per inflorescence and give the earliest flowering. In general, a planting density of 35,000-45,000 plants/ha is usually followed in Belgium, France and Italy. In Australia, a planting density of 35,000-40,000 plants/ha is followed for single beds and 1,72,000 plants/ha in 3row beds. In UK., a plant population of 50,000 plants/ha has been found profitable. In California and Florida, a population of 35,000-40,000 plants is followed for most of the cultivars. In India, a planting distance of 30 cm x 60 cm is considered best with a planting density of 55,000 to 60,000 plants/ha, though, research investigators have different views about the planting distance and population.

Planting time

Although planting can be done at any time between July and April, but early planting ensures a good crop. In temperate humid climate, planting is usually done in spring. If weather is undesirable, the runners can be stored in polybags at 0° C till conditions become favourable. Planting during July-August, however, needs adequate care, particularly from strong sunshine or dry soil conditions. During this period, irrigation should also be given frequently otherwise its leaves may desiccate and defoliate, which may result in delayed fruiting. However, planting in October and November usually does not require much special care for establishment of the plants. The planting season vary widely with the elevation of planting site and nature of the cultivars. For example, Chandler is planted during March under hilly tracts (Shimla) of India, whereas in sub-tropical zones, last week of October or first week of November is the ideal time for planting. If some cultivars are planted too early, their plants may lack vigour which results in poor fruit yield and inferior fruit quality. If planted too late, yield may be low but fruit size and quality are good. Similarly, if planted very late, runners develop in March bearing very less fruits. If planting is done very late (November or later than it), plants have very poor vegetative growth due to onset of winters in sub-tropical zones and thus have delayed flowering and poor yields.

Chapter 10

Water Management

Due to an evergreen, semi-herbaceous growth habit, rapid growth rate, high fruit productivity and relatively shallow root system, the growth and development of straberry is sensitive to variations in soil moisture. It responds to irrigation in a more profitable manner than any other fruit crop. Irrigation is also useful in utilizing the nutrients from soil, preventing blossoms from frost damage, hastening the growth of runners and increasing the yield and quality of the fruits.

Its vegetative growth is sensitive to moisture deficits. Generally, it gives optimum yield when soil moisture is 70-80 per cent of the field capacity. However, with the increased water use and soil moisture, dry-matter content increases but not the fruit yield and quality. Stolon production also increases with the irrigation, though, decreased stolon formation with the increased soil moisture has also been observed, possibly as a result of cooler soil and plant temperature or of excessive soil moisture with irrigation. Although flower and fruit number are generally dependent on adequate soil moisture, excessive irrigation may promote excessive vegetative growth, resulting in reduced flower-bud formation. In general, soil moisture deficits after flower emergence reduce flower number, fruit set, fruit weight, leaf number and leaf size and all these have a direct relation with fruit yield. Soil moisture deficit may also have adverse effect on fruit sugar, acidity and colour development. Thus, liberal supply of water during the time of growth and bearing is beneficial for strawberries, but excessive supply (flooding) should be avoided. When there is a fear of waterlogging, it is advisable to grow strawberries on raised beds or ridges.

Time of irrigation

Since strawberry is a shallow rooted crop and its roots are confined to the upper 15-20 cm of the soil surface, irrigation is required more frequently than any other fruit crop. Both shortage and excess of water supply should be avoided to harvest a good crop of quality fruits. Irrigation at the time of active vegetative growth and flowering is very necessary. It should, however, be avoided during ripening period because it may result in softening of the fruits. Such fruits are liable to be infested with fruit-rot. Moreover, the quality of fruits, especially the TSS, is adversely affected. The amount of water to be given also depends on physical condition of the soil, climatic condition and variety itself. The water should be applied slowly so as to avoid run-off. Excessive irrigation is, however, detrimental, as it may encourage growth of leaves and stolons at the expense of flowers and fruits. It may also result in the incidence of fruit-rot

The prevailing weather conditions and soil type also influence irrigation frequency. In case of sandy soil, frequent irrigation in summer is essential, especially if weather is warm and dry. Similarly, strawberries grown on heavy soils should only be irrigated once in two to three weeks during summer. The frequency should, however, be less in winter months.

Irrigation systems

Different systems of irrigation are used for irrigating strawberries. The most commonly used methods are the surface, drip or trickle system. These systems are used depending upon the soil type and economy of the farmers. Furrow system is commonly used if the subsoil is heavy and slopes are uniform. In the furrow system, water is applied between the rows and alleys are usually cultivated 2-3 days after each irrigation. Care must be taken to avoid the wetting of leaves since it may aggravate foliage scald. Sprinkler system of irrigation is best in localities, where there are more salts in the water and there is a problem of frost in spring. However, now-a-days drip or trickle irrigation systems are gaining popularity. This method is more convenient for regulating water as well as minimizing the amount of water. On an average, the amount of water used in trickle irrigation is 30-40% less, compared with the furrow or other methods of irrigation. Drip irrigation system has been used as a fertigation system also. The size, colour and quality of fruits improve greatly in trickle system of irrigation. Under drip irrigation system, the incidence of fruit- rot, the most dreaded disease of strawberry, is also reduced. In India, drip or trickle irrigation systems are common in strawberry plantations in Maharashtra. Recently some such units have been set up in Gurgaon, Hisar and Chandigarh.

Chapter 11

Nutritional Management

Strawberry also requires a number of mineral nutrients for proper growth and development. Practically all strawberry plantings benefit from the application of manures and fertilizers, but fertilizer is not a cure to all the problems, as assumed, because several factors like fertility and physical conditions of soil, careful planting, cultivars, adequate moisture, quantity, quality and time of applying the mulch, and weed and insect-pest control, may be as important as application of fertilizers. For example, certain cultivars are inherently more productive than others in a given region. Fertilizer application does not normally cause the yield of a low yielding or a disease susceptible cultivar to equal the yield of a high yielding and disease resistant cultivar. Similarly, fertilizer does not make a low-chilling cultivar to succeed in warm areas and vice-versa.

The essential nutrient elements required for strawberry are:

Nitrogen (N) : The N is necessary for proper growth of the plant. The N-deficient plants may have stunted growth and develop yellow-green colour. The leaves usually become small and the petioles look stiff and upright. Margins of older leaves develop bright orange-red colour and subsequently turn to brown and dry up. The runnering capacity and yield of plants is drastically reduced and if at all produced, they become bright red and thick. The N may also have some antagonistic effects on vegetative or reproductive growth of strawberry. Thus, excessive N may result in excessive vegetative growth and it may delay flower induction, and thereby the yield. Excessive nitrogen may also cause leaf injury, making the fruits softer and thereby affecting the fruit quality adversely.

Phosphorus (P) : Phosphorus, being a constituent of various nucleoproteins, enzymes and lipids, plays a vital role in the formation of new cells and promotion of root growth. The P-deficient plants have stunted growth. The leaves become usually dark green and curve downwards. Afterwards, the veins of older leaves turn bluish and the entire leaf looks bluish-purple. The affected plants have poor runnering capacity. The P is also considered as the best fertilizer to produce firm berries. In addition, P application may be effective in increasing the runner production, yield, berry size and weight. However, excessive P may result in aggravation of mottle leaf, low TSS and lower frost resistance. It may also render N unavailable, causing N deficiency in plants.

Potassium (K) : It takes part in some metabolic functions in growth and cell division of the young tissues. It is necessary for carbohydrate and protein synthesis. It enhances the ability of plants to resist diseases, insect-pest attack, cold, drought and other adverse conditions. Its deficiency causes reduction in growth, yield and fruit size, and tends to accelerate fruit ripening. Omission of potash fertilizer from a fertilizer schedule has serious effects on vigour, longevity and cropping of strawberry. Thus, increase in K level in plants may improve the fruit quality. However, its excessive application limits Ca and Mg absorption, which may lead to their deficiencies.

Calcium (Ca) : Calcium is an essential constituent of cell wall. It is apparently responsible for holding cells together. It is also an activator of phosphotase enzyme. It plays a vital role in carbohydrate translocation and better development of the root system. In addition, it neutralizes the toxic effects of Na, K and Mg in soil. Its deficiency results in burning of tips of young and unopened leaves, which subsequently become deformed and crinkled. The deficiency of Ca also interferes with anthocyanin synthesis, resulting in development of albinism, a disorder of strawberries. Further, fruits produced from Ca-deficient plants tend to rot more severely during storage. In general, small but continuous supply of Ca is essential for normal growth and fruit development and to minimize leaf-tip burn injury in the plants.

Magnesium (Mg) : It is the chief constituent of chlorophyll. It is also an enzyme activator, playing an important role in carbohydrate metabolism, phosphate transfer and decarboxylation reaction. It also helps in synthesis of organic acids in fruits. Its requirement is not high for strawberry but its deficiency sometimes occurs in gravel soils and when potasic fertilizers have been applied in excessive amount, which usually makes Mg unavailable to the plants. Its deficiency may result in yellowish-green colouration of leaf margins, followed by downward and inward curling of leaves. The basal portion of leaf along the main vein also turns yellowish-orange, while the areas adjacent to mid-rib remains green. Sometimes, Ca deficiency may result in poor runner production, delay in ripening and poor colour development in berries.

Zinc (Zn) : Zinc is involved in the biosynthesis of a plant hormone, indole acetic acid (IAA). It plays a vital role in nucleic acid, protein and chlorophyll synthesis. It also helps in utilization of P and N and acts as a catalyst in oxidation-reduction process. The deficiency of Zn results in appearance of pale-green to yellow leaves. The leaf blade becomes somewhat narrow and concave. The growth may be affected and may result in low yield of poor quality fruits. Fruit firmness and storability are also reduced. Excessive Zn may, however, cause Fe induced chlorosis and necrosis and retard growth adversely.

Boron (B) : An important constituent of enzymes and co-enzymes, B is required for proper development and differentiation of xylem and phloem. It is also involved in carbohydrate metabolism and functions in cell maturation. Plants deficient in B show a wide variety of symptoms, depending upon the species, cultivar and the age of plants but one of the earliest symptoms is failure of root tips to elongate normally. In general, B-deficient plants show reduced vegetative growth, producing mis-shapen or cat-faced fruits, which may be due to the poor achene development and subsequently, reduced tissue enlargement. Its deficiency also results in poor TSS accumulation and vitamin C content. However, B concentration beyond a certain limit becomes toxic to the plants. In general, leaves are good indicator of optimum B levels required in the plant. Its deficiency (< 25 ppm) and excess (> 200 ppm) induces leaf necrosis, resembling K deficiency symptoms. However, its supplementation enhances fruit set, fruit weight and yield.

Iron (Fe) : Though, Fe is not a constituent of chlorophyll, yet it is essential for its synthesis. It acts as a catalyst in many enzymes. The Fe-deficient plants are characterized by the development of a pronounced interveinal chlorosis similar to the one caused by Mg deficiency, but it occurs first on youngest leaf rather on older leaves as happens in Mg deficiency. Its deficiency causes leaf chlorosis and can greatly reduce yield if grown on high pH calcareous soils.

Copper (Cu) : Copper is a constituent of several enzymes, participating in cellular oxidation-reduction processes. Its deficiency first appears on younger leaves. The younger leaves of Cu-deficient plants turn pale-green with red areas between the larger veins. The symptoms become less distinct as the leaves grow older. Its deficiency is more severe if its contents are less than 3 mg/kg dry soil, which may adversely affect yield and quality of fruits.

Manganese (Mn) : It is essential for chlorophyll synthesis. It affects Fe transport, photosynthesis and N metabolism. It plays a vital role in chloroplast membrane system. The Mn deficiency in strawberry results in the development of pale-green to yellow colouration in the interveinal areas, while the veins remain green. Foliar spraying with $MnSO_4$ may correct its deficiency, though the concentration depends upon the deficiency limits.

Manures and fertilizers

It is essential to provide adequate manures and fertilizers for proper nourishment and to obtain uniformly high yield of quality fruits. However, it is very difficult to apply uniform fertilizer recommendation, acceptable to one and all as it depends on many factors like, soil type, fertility status of soil, planting system, climatic conditions, cultivar and the type of fertilizer to be given. Similarly, fertilizer doses are entirely different for open field cultivation and for protected cultivation. Further, strawberry may be planted for one year only (annual culture) and it may be perennial (three years or more). Therefore, manuring and fertilizer schedule should be decided accordingly. Normally, soil should be tested before planting and the manures and fertilizers should be applied accordingly. Usually, most of the macro- and micronutrients affect its growth and yield but N, P, K, B and lime should be supplemented in ample quantity.

Organic manuring : Organic manuring is perhaps the most wanted aspect in strawberry cultivation. Since its roots are confined to the top 25-30 cm layers of soil, the soil should be made friable by repeated ploughing and by supplementing with sufficient organic manures. Sufficient amount of organic manures (80 tonnes/ha) is recommended at the time of land preparation as they take more time for decomposition and to make the availability of nutrients to the plants. Similarly, adverse effects of the chemical fertilizers on soil can be counteracted if organic manures are applied with green manures like cowpeas, sunhemp, *dhaincha,* vetch, clovers etc. Soils containing more than 2.5% organic matter usually supply sufficient N for a good strawberry crop, but optimum quantity varies from location to location.

Usually, well-rotten farmyard manure (FYM) is the best for its cultivation. An application of 25-30 tonnes/acre of farmyard manure (FYM) prior to planting is the most ideal dose for strawberry cultivation. In addition, deep litter of poultry house, poultry manure and spent mushroom compost can also be used. These manures, however, contain more N, P and K content and hence the dose should be reduced. Thus, an application of 2-3 tonnes of poultry manure/acre is sufficient to supply the nutrients equivalent to 20-30 tonnes of farmyard manure. Thus, a pre-planting application of farmyard manure or other manure often provides all the needs of the crop over several years on a fertile soil. The availability of nutrients though depends upon the extent of roots, which in turn is decided by the depth of the soil and the availability of water in the soil.

For protected cultivation, different forms and doses of manures are recommended for its commercial cultivation. In greenhouses, it can be grown profitably on pumice mixed with sphagnum grass, mixture of peat and styromull (75:25), and even on broiler or layer litter. Similarly, rockwool, peat bales, and mushroom compost waste are other suitable media for growing strawberries under controlled (protected) conditions.

Fertilizer application : Depending upon different factors, different doses of fertilizers have been recommended. Different investigators have recommended different doses of manure and fertilizers, depending on locality, cultivar and type of fertilizer. However, an autumn application of 20 tonnes farmyard manure, 40 kg P_2O_5 and 20 Kg K_2O/ha, followed by spring application of 20 kg N after initiation of growth but before flowering is a common practice followed in India. The rate of K can be increased to 40 kg/ha in K-deficient soils, though Indian soils are not usually deficient in K. In the above recommendation, farmyard manure should be added as pre-planting dose during land preparation.

Foliar feeding : Foliar application of nutrients is a very effective tool for fertilizer supplementation because the leaves readily absorb the nutrients from the fertilizers. Generally, urea and some micronutrients can easily be applied through foliar means. Sometimes, even K and P can also be applied foliarly. Foliar application of nutrients may result in early floral initiations, more fruit set and yield per plant. Foliar application of N (0.5%), P_2O_5 (0.2%) and K_2O (0.5) during August-February is highly suitable for supplementing N, P and K respectively. Basfoliar combistrip (a micronutrient supplement), Fe EDDHA and $CaCl_2$ (0.5%) are usually applied through foliar means. Usually foliar sprays result in higher yield of better quality fruits in strawberry.

Tissue analysis : For predicting the nutrient status in plant, tissue analysis is considered as the most viable and reliable technique. In general, leaf blade is the best single plant part for indicating the nutrient status in strawberry. However, crown or leaflets are best for P, petiole for K, and roots for Mg. The recently matured leaf or leaves taken at the peak fruiting period may be taken for analysis. The critical levels and deficiency threshold of different elements at fruiting for N, P, K and Mg and in recently matured leaves for S, Fe, Mn, Zn, Cu, B and Mo are given in Table 4.

Table 4 : Critical limits and deficiency threshold levels of major and minor nutrients in strawberry

Nutrient element	Deficiency level	Critical limit
Major nutrients (% of dry matter)		
N	2.0	2.6-3.0
P	0.20	0.25-0.30
K	1.0	1.5·
Mg	0.10	0.15
Minor nutrients (ppm of dry matter)		
S	300-900	1000
Fe	5-40	50
Mn	4-25	30
Zn	6-10	20
B	<3.0	3
Mo	18-22	25
Cu	< 3.0	3.0

Fertigation : Application of fertilizers through drip irrigation system is an old fashion in strawberry gardens in the developed countries. In India, growing strawberry on commercial scale has started only recently. Poor financial position of the farmers and poor support of Government have forced strawberry farmers to adopt furrow method of irrigation. But in the recent past, a few microirrigation sets have been installed in some strawberry-growing areas of Maharashtra and

Haryana. Thus, farmers have shifted to applying fertilizers through drip irrigation units. The fertigation technique has some advantages over conventional fertilizer application methods. These are :

- Fertilizers are uniformly distributed and supplied at the required site.
- There is no wastage of fertilizers due to leaching or otherwise.
- Less labour is required.
- Transportation and storage losses are minimized.
- Herbicides for controlling weeds and pesticides (insecticides and fungicides etc.) can also be applied through drip or sprinkler microjet system along with fertilizers.

However, fertigation has some disadvantages like

- It requires higher initial cost.
- Technical skill is required.
- Local and cheap labour cannot be employed.
- If water is salty, it may damage the microjets. So, one has to be vigilant enough to check the pipes and microjets at regular intervals.

The decision on the amount of fertilizer to be used in fertigation is a job of a technical expert. It is really a very difficult task to decide the exact dose of macro and micronutrients to be supplied through drip irrigation because demand for nutrients depends upon many factors like soil type, climatic conditions, variety and type of fertilizer. However, in general, application of NPK in the ratio of 2:1.5:4 or 2:1.5:3 is good for optimum growth and higher yields of strawberry.

Chapter 12
Mulching

Mulching is recognized as one of the most beneficial intercultural practices in strawberry cultivation. It helps in better conservation of soil moisture, minimizes winter and frost injury, suppresses weed growth, reduces soil erosion and avoids contact of berries with soil, reducing the number of dirty and rot infected berries. As a winter protection, mulching checks heaving of plants and damage to roots caused by freezing and thawing. It also prevents the plants from drying out or freezing excessively. Sometimes, it makes picking more pleasant, may improve fruit colour and increases yield.

Characteristics of an ideal mulch

An ideal mulch material should be easily available, inexpensive, and free from weed seeds. It should not pack down easily. It should not cause any smothering effect on strawberry plants. It should not be too coarse or too light so as to blow away easily. It should decompose easily. It should not invite unwanted insect-pests and diseases and it should be a good incubator of heat.

Types of mulches

Different types of mulches are used in strawberry. A few of among them are:

Hay and straw : Clean wheat, rice or rye straw are the best mulch material for strawberry. Dry grass can also be used for mulch if it is available cheaply (Plate 12). About 15-20 tonnes of paddy straw is sufficient for mulching one-hectare field of strawberry. It is difficult to handle straw mulch under furrow-irrigated conditions than on drylands. In this condition, straw may be placed around plants by

Plate 11. Production of profuse runners by strawberry mother plant

Plate 12. Mulching strawberry with dry grass

leaving the space open between the rows for irrigation. If straw is not available in required quantity, it is advisable to grow some suitable crop for mulch material. Rye or sudan grass or a mixture of millet and sudan grass may produce a large amount of dry mulch material. Cotton burrs or broom sedge also make good and cheap mulch.

Before mulching with straw or hay, soil should be drenched with some insecticide and fungicide to protect the plants from insect or pathogen attack. Straw mulch, if combined with row covers, enhances the growth, yield and quality tremendously under protected cultivation. The spreading of straw by hand is a slow process but an ensilage spreader can do it fastly and cheaply. Hay and straw mulch material is easily available, and makes best use of waste material of cereals. It is comparatively cheaper than other mulch material (plastic mulch). It decomposes easily and adds manures to the field. Similarly, locally available labour can be utilised effectively for spreading the hay or straw. However, using straw as mulch material has certain disadvantages as straw may contain grain and weed seeds, which on germinating interfere with the main crop. It may contain harmful pathogens, which may infest strawberry plants and fruits also. Similarly, placing and removal of straw beneath the plants is a laborious and difficult task and new dressings are required every year.

Black plastic : Black plastic mulch is used commercially in different countries. Black plastic of 200-400 gauge with widths of 3, 4, 6 ft or more with desired length is used. The plastic may be laid over the beds before planting or after 1-4 weeks of transplanting as per the desire of the grower. Many growers in some developed countries make the beds first and lay plastic with the help of a machine. Planting is then done through the plastic by hand. However, other growers apply and slit the plastic and pull the plants through it.

Using black plastic mulch has certain advantages, as the plants grown under black mulch are more vigorous. The black plastic mulch can be recycled time and again but not hay or straw. Usually higher yields of better quality fruits are obtained. Berries ripen early and rotting of fruits is reduced greatly. It ensures more constant moisture conservation thereby reducing irrigation frequency. It checks leaching of nutrients and avoids washing down of beds. Cleaner fruits are obtained with lesser fruit cracking during wet periods. However, its disadvantages are that plastic mulch is costlier than hay or straw, it does not decompose easily, it absorbs heat easily and fruits coming in contact with mulch develop unwanted and unattractive spots. Moreover, it has to be removed every year.

Clear plastic mulch : Clear plastic of 200-400 gauge is also used as mulch in strawberry. It is mostly used in winter plantings. Clear plastic mulch is usually applied a few weeks after planting. Clear plastic mulch may be helpful in tremendous increase in the yield. This is primarily due to the temperature effect. Since the success of winter plantings depends on how much plants grow in winter, the use of mulch becomes necessary. If mulch is applied too late, it reduces the yield by stimulating runner production. White plastic sheets are also used as mulch material in strawberry.

Coloured mulching films : In addition to black and clear plastic, some coloured PVC films like red, violet, grey, green and brown are also used as mulch material in strawberry plantations.

Biodegradable papers : Biodegradable paper with a polyethylene coating on both sides or with a single polyethylene coating is also used as mulch in strawberry plantations. These mulches also ensure good fruit yield, comparable to black polyethylene mulch. The added advantage of biodegradable paper mulch is that it need not to be removed from the field at season's end because it decomposes easily. These are, however, very costly compared to black or white polyethylene sheets.

Pine needles: Pine needles act as an excellent alternative for their use as mulch material in strawberries in mid-hills of Himachal Pradesh, where these are produced in plenty in the thick forests of pine trees. About 10 tonnes of pine needles are required for mulching one hectare strawberry field.

Time for mulching

The best time for mulching in the hills is determined by the prevailing temperature. Mulching should be done in winter when the temperature has come down to 2-3° C or so. The ideal period for mulching is rather short and varies from year to year. In general, mulch is usually applied in December or January, depending on the area, plant size, planting date and temperature requirement of a given cultivar. If plants are under-developed or planted later than the recommended date, apply mulch earlier, possibly even in November, as soon as runnering has ceased. But do not apply it early if planting is done on proper time since yield, fruit size and quality are affected unfavourably. In fall bearing cultivars, if full winter protection is not provided, apply straw mulch several weeks before bloom to protect berries from dirt which helps them to be cleaner and brighter. Scatter the material over the beds, completely covering the plants. But if mulch is to be applied after the plants have started to bloom, care should be taken not to cover the flowers and developing berries.

Depth of mulch

If straw is used as mulch, apply clear straw several inches deep both over and between the rows. About 5 cm to 7.5 cm deep mulch, when settled, requires about 8-10 tonnes straw/ha. A clay soil subject to heaving needs slightly heavier mulch than a lighter soil. Too little mulch may shorten the picking period, resulting in small berries as well as weakening of the roots where winters are milder. About 5 to 6 tonnes of bale straw mulch/ha is a good amount. This should be applied in a such way that plants are visible through the mulch. Too heavy mulch in mild winters may also result in smothering effect. Similarly, heavy covering of straw is conducive to ice formation in the mulch as a result of which, ice layer tends to lower the temperature under the mulch and by doing so may defeat the purpose of mulch as winter protection. To keep the mulch from blowing away in strong winds, long poles may be placed on top of it down the row. The plastic mulches are applied deeply but the ends are buried deep in the soil.

Removing mulch

Remove the winter mulch in early spring after the danger of severe frost is over. If mulch is removed too early, flowering may begin early, which may be destroyed by the frost. Similarly, late removal of mulch may produce small-sized fruits, enhancing the incidence of several leaf diseases. If covering is thin, the plants may grow up through it, though some shaking of straw is advisable. If mulch is heavy, part it over row with a fork and push it into the alleys between rows, tramp it down, and leave it there until harvesting. However, mulch should be removed earlier from the propagation rows than from the fruiting plants. Where late ripening is desired, allow mulch to remain on plants somewhat later than usual. With this plan, the leaves may become yellowish before the mulch is lifted. Though this is not a desirable practice, however, it delays the picking season. In India, mulch material (paddy, wheat straw etc.) is not removed until harvesting and is allowed to decompose in the field, which acts as an organic manure. The plastic mulch material is, however, removed every year. In developed countries, wind row-turning equipment and side delivery rakes are used for mulch removal.

Effects of mulching

Mulching has strong influence on growth, yield, quality and duration of harvesting in strawberry, which is primarily due to better soil and moisture conservation, changes in soil temperature, improved nutrient availability by suppressing weed number and growth, protection from frost injury and by way of reducing the number of dirty and diseased berries. Thus, using a suitable mulch at a suitable time can optimize economic returns from strawberry. Mulching is, therefore, considered as the most beneficial cultural practice in strawberry. The effects of mulching are:

Modifications in soil temperature : Change in soil moisture temperature is the most significant influence of mulching. About 70 per cent or more yield in mulched winter strawberry plantations is obtained primarily due to temperature effect. Irrespective of mulch material used, there is rise in soil temperature by 1-3° C. The rise in temperature is more with coloured plastic sheets compared to translucent sheets, paddy and wheat straw mulch. Black polyethylene mulch may raise the soil temperature by 3-5° C compared to un-mulched fields. Temperature also increases with white plastic mulch but the rise is comparatively less than the black polyethylene mulch. Significant improvement in soil temperature provides appropriate environment for growth, development and flowering of strawberry plants, resulting in increased yield and quality.

Plant growth : High yield is obtained if plants are healthy. Mulching aids in proper growth and development of strawberry plants by modifying soil temperature, and providing better nutrient availability and moisture conservation. Mulch also provides better environment for proper growth and development of roots.

Runner production : Production of runners usually increases with different mulch materials. It is due to availability of better growth conditions under mulch. In general, runner production is higher with coloured mulch compared to white plastic mulch or the other materials like paddy or wheat straw.

Moisture conservation : Mulches are efficient tool for better moisture conservation by way of reduced evapo-transpiration rate. Thus, amount and frequency of water or irrigation is reduced if suitable mulch is used timely. All mulches increase total water content, thereby reducing the irrigation frequency.

Nutrient uptake : Mulch suppresses weed growth considerably, making availability of more nutrients to the strawberry plants. Further, due to better root growth and moisture conservation under mulch cover, nutrient uptake is also better, which helps in better growth of the plants.

Weed population and growth : The use of mulch in strawberry has rigid control over growth and population of weeds. The weeds may, however, grow to some extent in paddy straw but their growth is lanky. However, weed growth is completely suppressed under polyethylene mulches. The use of black polyethylene as mulch material is highly economical because it reduces weed growth and population to a greater extent, thereby reducing labour requirement for manual hoeing and weeding.

Flowering and fruiting : Mulch may advance flowering and fruiting in strawberry. Its effect, however, depends on type of mulch, cultivar and time of application of mulch. Mulching with clear polyethylene in winter may advance flowering by a week or so, whereas, straw mulch increases number of primary, secondary and tertiary flowers and advance flowering by 10-15 days. Early flowering and fruiting in strawberry is associated with increase in the soil temperature and prevention of flower buds from frost damage.

Fruit yield and quality : Mulching increases yield and improves quality of the fruits. The yields may increase in winter plantations with clear plastic mulch up to 70 per cent or more compared with no mulch, which is primarily due to the rise in soil temperature and better growth of the plants. However, increase in production in summer plantations due to plastic mulch may not be as high as winter plantings, but it does modify the production pattern favourably, increasing fruit size, yield and quality of fruits. The yield of strawberry during summer may increase by 50 per cent by mulching. Mulching not only increases fruit yield but also enhances fruit quality and colour of the strawberries invariably.

Incidence of diseases : Fruit-rot (grey mould) is a serious disease of strawberry. It is usually more prevalent if berries come in contact with the soil. Mulch avoids direct contact of berries with soil, reducing the number of dirty and rot infected fruits.

Chapter 13

Management of Weeds

Weeds are unwanted plants responsible for substantial loss in yields by competing for space, moisture, nutrients and solar energy. They also hinder many cultural operations like harvesting and act as hosts for many insect-pests and diseases. Thus, their management is of paramount importance. Different kinds of weeds have been recorded in strawberry. Since its cultivation is of recent origin in India, not much is known about the occurrence of different weeds in strawberry fields. However, *Portulaca oleracea, Amaranthus retrofleus, Chenopodium album, Agropyron repens,* doob grass *(Cynodon dactylon), Ranunclulus arvensis, Abutilon, Trifolium, Parthenium,* and *Cyprus rotundus* are common weeds in strawberry fields. In Germany, *Matricaria chamomilla* and *Capsella bursa-pastoris repens* are the main weeds found in strawberry fields. The *Ranunculus repens* and *Trifolium* are the weeds chiefly found in Irish Republic. In the Netherlands, *Rorippa silvestris* is considered as one of the most destructive weeds in strawberry fields. Similarly, *Chenopodium album, Amaranthus* and *Cyprus rodundus* are common weeds in strawberry fields in Greece. In USA and Canada, some grasses like *Viola arvensis* and *Elynus repens* are important weeds in strawberry fields. Some common grasses found in strawberry fields are crabgrass (*Digitaria sanguinalis*), stinkgrass (*Eragrostis ciliensis)*, foxtail grass (*Setaria* spp.) and *Panicum dicholomiflorum* .

Weed control is an essential intercultural operation for a healthy crop. Prevention, eradication and control of weeds can reduce the losses caused by the weeds.

Weed prevention : To avoid the entry of weeds by maintaining proper sanitation is of vital importance. It can be achieved by cleaning the contaminated tools and implements, and avoiding use of mulch material contaminated with weed seeds. Preferably polyethylene should be used as a mulch material because the paddy straw or other cereal straw may contain seeds of unwanted plants (weeds).

Weed eradication : Complete destruction or removal of weeds by any means is known as weed eradication. Eradication of perennial weeds like *doob (Cynodon dactylon)* and *Cyprus rotundus* is particularly very necessary, as they cause heavy losses to strawberry.

Weed control : Owing to low growth habit, strawberry cannot compete successfully like other fruit crops with weeds that deprive plants of moisture, sunshine and nutrients. Therefore, weed control is considered as one of the most expensive management practices in strawberry. In most cases, prevention and eradication of weeds is not easy and feasible. Therefore, their control by one or the other means becomes necessary. Weeds in strawberry field can be controlled by different methods, viz. mechanical, crop rotation, biological means and by use of different chemicals. These methods can either be used separately or in combination with other methods for desirable results.

Mechanical control

Hoeing and hand weeding are most effective ways to keep weeds under control. Deep cultivation or repeated cultivation of land before planting can also keep most of the weeds under control. Several hand weedings may be needed during the first year or before picking next year. It is important to keep fields free from weeds during later summer, especially when fruit buds are forming. It is better to place weeds under developing fruits after their removal from the field. It serves as good mulch for better soil moisture conservation. Moreover, it avoids direct contact of fruits with soil, resulting in low fruit-rot incidence. Some special machines or tools like shovels, hoes duck foot or blades etc. are also used in developed countries for weed control. Mulch also suppresses the growth and further spread of weeds to a greater extent. Polyethylene sheets (black and white) are more effective in controlling weeds compared to paddy straw mulch. Even, cloches have been found to be effective in controlling weeds in strawberry fields.

Crop rotation

A few weeds may grow in fields where strawberries are planted in a rotation. Generally, new crop competes successfully with weeds of previous crop. In strawberry, three cultural systems, viz. rice and strawberry, maize and strawberry and grape and strawberry have been proved to be highly beneficial in reducing the weed population.

Biological control

With the advancement in agro-techniques, use of biological agents (insects, fungi and birds) to control weeds has also been used as an effective tool. Herbicides are mostly used to control a wide range of weeds but lack of tolerance of crops and soil residue etc. cause many problems and thus use of bio-herbicides, plant pathogenic organisms (fungi), insects and birds etc. to control weeds is gaining popularity day-by-day. Among birds, geese is quite effective as it eats away grassy weeds without damaging the fruits. These birds are particularly effective in during rains they are able to move about in the field when man and machine cannot. However, geese does not eat broad-leaved weeds but can be trained to do so. Goslings are more efficient than the grown up geese. These birds feed mostly on weeds during early morning hours, late afternoon, evening or even on moonlit nights. They seldom roam the berry field in the heat of the day. Since, geese prefer young tender growth, bring them soon after planting or when grass starts appearing. During fruiting, remove the birds from the field during the bloom and berry periods and again bring back after picking is over. Usually, four to five geese/ acre are sufficient in a normal growing season. However, in wet season, six to eight geese/acre are required for keeping the population of grassy weeds under control. It is advisable not to use herbicides or insecticides if geese are to be kept in field for the purpose of weed control.

Some fungi like *Colletotrichum* sp., *Alternaria* spp. and *Cercospora* spp. also control some weeds by way of infestation and destruction by causing diseases in them. Some insects, particular beetles also destroy weeds by feeding on leaves. A list of some biological agents used in controlling weeds in strawberry field is given in Table 5.

Table 5: Biological agents for weed control in strawberry

Biological agent	Weed controlled
1. Insects	
Beetle (*Zygogramma biocolorata*)	Congress grass (*Parthenium* sp.)
2. Fungi	
a. *Colletotrichum malvarum* (controls weeds by inducing anthracnose disease in them)	*Sida spinosa*
b. *Colletotrichum coccoides*	*Abuliton theophrasti*
c. *Alternaria cassiae* (controls weeds by inducing blight disease in them)	*Cassia obtusifolia*
d. *Cercospora rodmanii* (controls weeds by inducing leaf spot disease in them)	*Eichornia crassipes*
3. Birds	
Geese	Most of the grassy weeds

Chemical control

In general, control of weeds through mechanical means is not possible because it is quite labourious and cumbersome. However, mechanical control should be supplemented with herbicidal control. Thus, chemicals are useful in weed control, reducing the amount of hand labour and number of cultivations. Herbicides are mainly used in three different stages of weed control in strawberry fields. They are, perennial weed control before planting, annual weed control at planting or in young plants, and annual weed control in an established crop. For this, weedicides are usually applied in soil or are used as foliar sprays.

1. **Foliar application** : Foliar application of herbicides is done on the growing weeds after their emergence in the field. Weeds are killed by the herbicidal action either by contact or by their translocation into weed parts. Contact herbicides are most useful for controlling weed seedlings, whereas herbicides having translocated action can be used in perennial as well as on annual weeds because they kill the entire plant.
2. **Soil application** : Soil application of herbicides is done at pre-planting time for controlling weeds which are still to emerge (pre-emergence). These herbicides may be applied to soil surface or may be incorporated into soil by cultivation, fumigation or may be injected below the soil surface by one or the other means.

Herbicidal recommendations

Strawberry is a herbaceous fruit crop and therefore utmost care should be taken while using herbicides. Only optimum doses of the recommended herbicides should be used as some herbicides may cause some phytotoxic effects like scorching of leaves, veinal chlorosis and necrosis of leaf tissues. At the same time, the herbicidal recommendations cannot be uniform for a given place as the development of new chemicals (herbicides), soil type, climatic conditions, type of weed, application time etc., change the picture from time to time and recommendations may quickly become outdated. In general, most of the annual weeds can be controlled with herbicides in newly planted strawberry beds and to control germinating annual broad-leaved weeds, herbicides should be applied within two weeks of plantings.

Different recommendations of herbicidal use in controlling weeds in strawberry field have come up in different countries. In general, alachlor (2 kg/ha), venzar (1.5-2.0 kg/ha) or gramoxone (3.0 l/ha) or treflan (1.5 kg/ha) can be used as pre-planting herbicides to control weeds. Post-emergence application of napropamide provides good control to broad-leaved and annual grassy weeds in strawberries. Simazine (5 kg/ha) in autumn is quite effective in controlling broad-leaved weeds. Stomp also controls wide range of annual weeds in strawberry. Similarly, chloroxuron is also a well-known herbicide for post-emergence control of annual broad-leaved weeds. About 90 per cent of broad-leaved weeds in strawberry can be controlled with chloroxuron. Fumigation with methyl bromide and other fumigants before planting is equally effective in controlling weeds and nematodes as well. In California, soil fumigation with methyl bromide and chloropicrin mixture under polyethylene mulch is a commonly used practice for control of weeds, nematodes and soil-borne diseases. In general, a herbicidal schedule given in Table 6 may be followed for full control of weeds in strawberry plantations.

Table 6 : Some herbicidal recommendations for weed control in strawberry

Planting type	Time of application	Herbicide	Rate (a.i./ ha/ 800-1000 l water)	Remarks
1. New plantings	One or two days before planting	Paraquat and Diaquat	1 to 2 Kg	Use these herbicides only if it is necessary to destroy the weed seedlings.
	Shortly after planting	Lenacil or Chloroxuron or Simazine	4 to 5 Kg 0.5 to 0.7 Kg 0.5 to 0.7 Kg	Use only Simazine if soil conditions are known.
	Late October to early December	Simazine or Lenacil	0.5 to 0.7 Kg 2 Kg	Use these herbicides if planting has to be done in August or early September.
	March-April (before opening of flowers)	Chloroxuron or Lenacil	4.0 to 5.0 Kg 2.0 Kg	Do not use these herbicides if Simazine has already suppressed weed growth.
2. Established plants	August–September	Simazine	0.7 Kg	Apply these herbicides after mechanical control.
	November-December	Simazine or Lenacil	0.7 Kg 2.0 Kg	Apply these herbicides as pre-emergence weedicide.
	March-April (before opening of flowers)	Simazine and Lenacil	0.7 Kg 2.0 Kg	Do not apply these herbicide if winter application has suppressed the weeds.
3. Runner beds	End of June-July	Simazine or Linacil	0.5 to 0.7 Kg	Apply them at the time of the appearance of runners.

Chapter 14

Protected and Hydropic Cultivation

Protected cultivation : In recent years, great emphasis has been laid on protected cultivation of strawberry. In fact, growing strawberries out of the normal season and locality has led to the emergence of the idea of protected cultivation. Protected cultivation basically involves the use of plastic greenhouses for forcing and prolonging the harvesting season by protecting the plants from the vaggaries of adverse weather conditions. Protected cultivation is now becoming very popular in a number of countries. Interestingly, strawberry breeders world over, have oriented their research programmes for breeding infra short-day varieties which are suitable for greenhouse conditions. Similarly, hydroponic culture system (soilless culture) for strawberries in greenhouses is also becoming popular in some advanced countries.

The technique of protected cultivation is very simple and easy to operate. It involves the construction either of plastic greenhouses or polyethylene tunnels or row covers of plastic film over wire hoops or open tunnels or cloches at a particular time of the year. Generally, dwarf varieties grow better under protected conditions. Some varieties like Elsanta, Rosella, Gorella, Brighton, Fern, Selva, Lassen, Shenxu, Douglas and Chandler grow better under protected cultivation. In addition to CO_2 enrichment (750 ppm) inside greenhouse, much attention should be paid on nutrition and light requirement. Most benefits of protected cultivation result from the increase in the mean temperature during winter, which not only save the plants from the danger of frost but also help in proper growth and development of plant, early flowering and fruiting in strawberry.

The medium for growing strawberries in greenhouses is soil but now there has been a strong shift towards other substrates like rockwool, peat bales and spun fibre etc.

Effects of protected cultivation

Some of the beneficial effects of protected cultivation are:

Plant growth and development : Plants grown in greenhouses, plastic tunnels, cloches or row covers have better growth and lush green foliage. Covered plants usually become photosynthetically active and independent much earlier than the uncovered plants. The average rise in mean temperature in spring and longer exposure to light in greenhouses favours high photosynthetic activity in covered plants, resulting in proper growth and development of the plants.

Frost injury : Winter injury is a great limiting factor in commercial strawberry cultivation. There is no danger of frost injury in protected cultivation. The protection from the frost is mainly due to rise in temperature under the protected structures. Sometimes, temperature inside the row covers and low plastic tunnels rises excessively but the perforated plastic tunnels never allow the temperature to go beyond 35° C. Moreover, these tunnels offer better ventilation compared to low un-perforated plastic tunnels.

Incidence of soil-borne diseases : The incidence of soil-borne diseases is very less in protected cultivated strawberries. Soil disinfection by solar heating or the use of sterilized media like rockwool, peak bales etc., avoid the chances of infection by soil-borne diseases.

Runner production : The production of runners is enhanced when row covers plastic films over wire hoops are used during winter. It is primarily due to better growth of mother plant and favourable environmental conditions for the development of runnners.

Crop maturity : The flowering and fruiting of strawberry is enhanced invariably under greenhouse. The flowering in plastic covered strawberries may be advanced by 20-25 days and fruiting by 15-20 days. With the regulation of temperature in protected structures, the availability of strawberry can be staggered at least for a month or so, which is otherwise not possible under open fields.

Fruit yield : The fruit yield of different cultivars increases substantially under greenhouse, row covers, perforated tunnels and cloches compared to open fields. The increase in yield is primarily due to the increase in the development of secondary and tertiary berries under protected structures.

Fruit quality : The quality of berries produced under protected structures is better than the fruits produced in open fields. The berries produced under protected structures have usually better colour development, higher TSS, better flavour and low acidity as compared to those produced under open fields.

Hydropic cultivation

In advanced countries, the emphasis is laid on growing strawberry hydroponically in greenhouses. In hydroponic system, plants are planted in a nutrient filled moveable gullies suspended from the greenhouse roof. Different substrates like perlite, peat, pumice etc. may be used as nutrient medium in this system but usually perlite and peat mixture give better results. The soilless strawberry production is ideal for "off-season" fruiting.

In India, protected and hydroponic culture systems are not followed in strawberry. It is usually grown in open fields in hills, northern plains under sub-tropical climate and in some areas in tropical climate. Plants produce frequent runners in hills but severe hot conditions after harvesting limit runner production in the plains. Now, some farmers have initiated construction of low-cost polyhouses, polyethylene tunnels and cloches in plains and hills during winter to prevent the plants from frost injury and to get early crop, but such protected cultivation is not followed commercially as in other countries.

Chapter 15

Flowering, pollination and fruit set

Flowering

Light is the major environmental factor, which regulates growth and development of different species of *Fragaria*. Variation in irradiance level may invariably affect leaf size, root and crown development, stamen development, stolon formation, fruit set, fruit size and yield. The quality of light can also affect physiology of strawberry. Under floral inductive conditions (8-hr days), flower initiation in strawberry is usually delayed or inhibited with far red light (700-800 nm). Similarly, stolon formation, leaf lamina and petiole length all increase with increased exposure to far red light during first half of the each day's dark period or with the exposure to red light during the second half of the dark period. Strawberry flower is a perfect flower having five sepals, five petals and numerous stamens (Plate 13). Upon pollination, it produces aggregate fruit with numerous seeds.

The flowering, pollination and fruit set in strawberry is basially influenced by photoperiod, temperature and photoperiod x temperature interactions as under:

Effects of photoperiod

The famous scientist Sudds was the first to report the photoperiodic effect on strawberries in 1928. Accordingly, the strawberry cultivars have been categorized as short day (SD), long day (LD) and day neutrals (DN) on the basis of photoperiodic requirements for floral initiation. In addition to floral induction, leaf size, petiole length, and stolon development are also highly sensitive to photoperiod. The short day (SD) cultivars are commonly called as June-bearing cultivars, in which floral induction occurs with photoperiods of less than 14 hours. Though, most of short-day types show a facultative short-day response, in which floral induction may occur more or less continuously regardless of the day length, provided that temperature is less than 16° C. Although, short-day cultivars may have different critical photoperiods, most of them bloom independently of day length with moderately cool temperature. The day-neutrals (DN) are called as ever-bearing (EB) cultivars that generally flower continuously regardless of the day length. Although temperature also modifies photoperiodic response of day-neutral cultivars but they are less sensitive than the short-day cultivars. The group of long-day (LD) type is between ever-bearing and day neutral types, which requires long day conditions for floral initiation. Though, different cultivars in all three categories are strongly influenced by photoperiodic influence but major difference among photoperiodic groups is a decreasing sensitivity to high temperatures from short-day to ever-bearing, from ever-bearing to day-neutral group. Thus, flowering in strawberry ranges in a continuous gradient from obligate single cropping to facultative short day, through the various degrees of everbearing or day-neutral types.

In general, flowering in strawberry occurs under short light period (10 hr) with long dark period (14 hr), short light and dark periods (10 hr), and long light and dark periods (14 hr). Thus, duration of dark period rather than light period is the factor, which controls floral initiation in strawberry. Photoperiod has a marked effect on its vegetative growth, morphology, yield, stolon formation, petiole length, leaf area etc. Yield increases with the increasing photoperiod, though the effects are cultivar-specific.

Effects of temperature

The growth and development processes in strawberry are highly sensitive to variations in air and soil temperature. Simple effects of temperature are often cultivar and species dependent. The growing season temperature of 15° C is optimum for most of the cultivars, though some cultivars grow best even at 25° C. Some research workers have reported a temperature range between 20° C and 26°C to be the most ideal range for proper growth and flowering in strawberry, though the temperature effects are also cultivar specific. The optimum soil temperature for its vegetative growth appears to vary with the culture. For example, for Shasta and Lassen cultivars, a soil temperature of 12.8 °C and 7.2° C is the best. The fruiting in both these cultivars occur at 12.8° C and 7.2° C, while stolon development takes place at or above the soil temperature of 18.3° C. In general, optimum temperature between onset of flowering and ripening is 13.8° C for early cultivars, 14.4° C for mid-season and 15.1° C for late cultivars. Temperature below 15.6° C may inhibit pollen germination and pollen tube growth, resulting in misshapen fruits. Temperature can also affect fruit set and fruit quality due to indirect effect on bee activity. Although cultivated strawberry has perfect flowers and is self-fruitful, pollination is enhanced by bees but the bee activity is hindered at temperatures below 10° C and above 35° C. Thus, cool temperature, rain, or strong winds, inhibit bee flights, resulting in decreased pollination and increased development of misshapen fruits.

Strawberry species and cultivars vary in their ability to withstand low temperature. Although certain cultivars withstand even a winter temperature of –50° C but strawberry is not as cold hardy as most temperate fruit crops. Freezing temperature frequently injures the strawberries. The injury, however, depends on cultivar, stage of plant development, temperature, duration of temperature, pre-conditioning, climatic conditions, plant nutritional status, soil and plant moisture status, rate of freezing and thawing, mulching etc.

In autumn, exposure of plants or plant parts to cool temperature and short photoperiod results in floral induction, branch crown development, reduced leaf size, and with prolonged exposure to the onset of dormancy. Unlike most temperate fruits, dormancy in strawberry is controlled solely by external conditions (ectodormancy), such as temperature and photoperiod, dormant (resting) plants retain green leaves and can resume growth when environmental conditions are favourable. It means there is no true dormancy in strawberry but in absence of sufficient chilling, dormant strawberry plants have low compact growth and yield poorly and thus sufficient chilling should be given before planting strawberry runners.

The amount of chilling required to terminate dormancy (rest) in strawberry is generally cultivar dependent. Generally, temperate zone cultivars require longer period of chilling to overcome dormancy, whereas, subtropical cultivars require little or no chilling period. The duration of chilling period required for breaking rest period, ranges from as little as two to four weeks for Tioga and Torrey, to eight weeks or more for Redgauntlet and Hayward 17.

Spring or autumn frost often results in injury to its reproductive organs. Cold injury to open strawberry flowers often occurs at temperature of about -2° C, although flowers of several cultivars even survive even at -4.5° C. Some species of *Fragaria* can even tolerate a temperature below -50° C. The susceptibility to cold may vary with cultivar and degree of ice nucleation that occurs in tissues.

Effects of photoperiod and temperature interaction

The photoperiod and temperature interaction plays a pivotal role in regulation of vegetative and reproductive development in strawberry. Under short-day conditions, floral initiation is usually maximum at 21° C, intermediate at 15.6° C and lowest at 12.8° C. With a 16-hour photoperiod cycle, optimum temperature for floral initiation is 15.6° C. But, in general, cultivars differ widely with regard to floral initiation in response to photoperiod and temperature interaction. For example, under long day conditions (14 to 16 hrs) and at 21° C, Klondike produces only stolons, but Burrill can produce equal number of stolons and flower clusters, whereas at 15.6°C, Klondike produces both flowers and stolons but Burrill does not.

Although the optimal day length for floral induction in short-day cultivars appears to be between 8 and 11 hrs, the role of photoperiod is more critical at temperature above 15° C. Thus, critical photoperiod for floral induction is variously reported as between 12 and 16 hrs; 12 and 14 hrs; 11 and 14 hrs; 11 and 15 hrs; 13 and 15 hrs or 13 and 16 hrs, depending on cultivar, temperature and photoperiod interactions.

The fruiting season for short-day cultivars tends to be longer in areas with mild winter due to longer growing season in which photoperiods are less than 14 hours and chilling injuries are also few. In general, critical photoperiod for inhibition of flower induction and for promotion of vegetative growth, such as increased petiole length, leaf size, and stolon formation are similar. However, once the floral induction and initiation have occurred, further development is enhanced by exposure to long day conditions. The number of short day photo-inductive cycles required for floral initiation in short-day cultivars is highly variable due to temperature and photoperiod interactions, as well as seasonal and pre-treatment factors. The minimum number of short-day cycles required for floral induction is proportional to the temperature and usually it ranges between seven and fifteen but at as many as 24 or more cycles may be there but these cycles are still higher at higher temperature. For example, with a 16-hr photoperiod, more than 16 cycles are generally required for floral induction at 17° C but only 10 cycles are needed at 9° C. Similarly, with 8-hr photoperiod, only 10 cycles are required for floral induction at 24° C but more than 20 cycles are required at 30° C.

Pollination

Most of the strawberry cultivars produce hermaphrodite flowers and are self-fertile. However, some species also produce male or staminate, imperfect and female or pistillate flowers. The hermaphrodite flowers are self-fertile but pistillate flowers require cross-pollination for fruitful production. In some European varieties like Korona, Gorella and Confitura, partial sterility has also been reported. In general, first flush of flowers in perfect flowered varieties open and set well, but the later flowers set fruit partially or do not set at all. In these varieties, natural sterility is the primary cause of poor fruit set. However, sometimes, the first flushes of flowers develop into *nubbins*, while later flushes produce good sized berries. The poor development of berries into *nubbins* is probably due to partial pollination as sometimes petals and sepals may cover the pistils of flowers and thereby prevent pollination. The frost and insect injury may also produce *nubbins*. Similarly, poor fruit development in some varieties may be due to the fact that they produce either pistillate flowers only, flowers with few stamens or stamens that fail to produce sufficient pollen for fertilization. Further, plants of same variety vary in their capacity to produce pollen from one area to another. The first to open flowers in the spring in many varieties may develop good anthers with abundant pollen but almost none in other years. Thus, a variety may be self-fertile in a particular locality but self-sterile in the other. For example, Crescent and Stirling Castle cultivars produce only pistillate flowers in USA and require suitable pollinator and pollinizer for fruit set, but they produce hermaphrodite flowers in England and are self-fertile, requiring no pollinator or pollinizer.

The presence of pollinators and pollinizers in the vicinity of strawberry plantings is important in cultivars whether they are self-fertile or self-sterile. In self-sterile cultivars, fruit set is improved considerably as cross-pollination is required for them. In absence of a suitable pollinator and pollinizer, or insufficient pollination, the percentage of malformed or misshapen fruits may increase. Therefore, provision of a suitable pollinator along with planting of a suitable pollinizer with the main cultivar is important for fruitful yield. Generally, pollinizer (one row) and main cultivar (5 rows) is necessary for commercial plantations.

Honey-bees are the chief insect pollinators of the strawberries. The other insect pollinators are blowflies and bumble-bees. At least 3-4 beehives/ha should be placed in its field for proper pollination. The placement of honey bee colonies improves fruit set, yield and quality and reduces the number of malformed fruits. The bee activity is usually hindered at a temperature above 40° C or below 4° C. Bees take active flights at 10-15° C. Similarly, bees do not take flights in rainy weather and in strong winds. Thus, climatic conditions also hinder pollination and fruit set in strawberry by interfering with bee activity. In absence of insect pollinators, wind also plays an important role in pollination of strawberry. However, strong, hot and desiccating winds may dry the stigmatic fluid of the flowers, which may affect fruit set. Similarly, hot and desiccating winds may favour the drop of developing fruits.

Fruit set

Fruit set is the characteristic feature of a cultivar that is influenced by management practices, photoperiod, existing temperature, other climatic factors and activity of insect pollinators. In general, its inflorescence has one primary, two secondary, four tertiary and eight quaternary flowers. But maximum fruits are produced from the tertiary flowers. Even under proper management, most of the varieties do not give cent per cent fruit set due to one or the other reasons. The fruits can however, be increased to a limited extent by way of application of nitrogenous fertilizers, auxins and with the provision of insect pollinators, especially honey bees, in the vicinity of its plantations.

Chapter 16

Fruit growth, Development and Ripening

Fruit Structure

After adequate pollination and subsequent fertilization, ovules develop rapidly and the flesh around each fertile seed starts swelling. It develops into an unusual structure as a false fruit, having true fruits or achenes on the outside of a fleshy receptacle, attached to it through a vascular connection. Strawberry is an aggregate fruit, which is developed by the simultaneous ripening of a number of separate berries of a single flower, adhering as a common unit on a common receptacle (Figure 5).

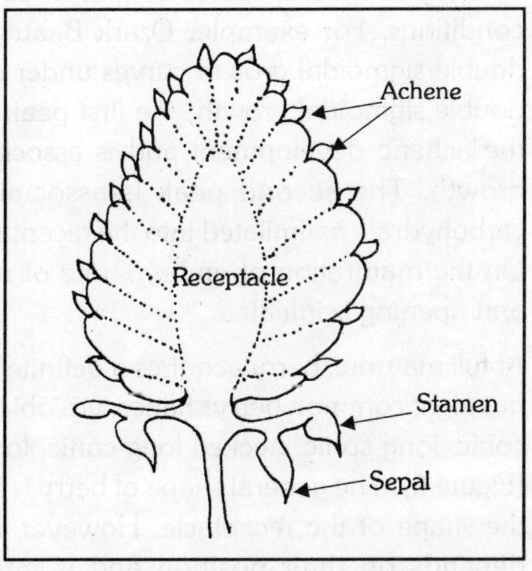

Figure 5. Structure of the strawberry fruit

Botanists called the strawberry fruit as an etaerio of achenes. Usually, primary berry in a cluster is the largest fruit, whereas, secondary, tertiary and quaternary are progressively smaller in size. Each berry may have 20-500 seeds depending on the cultivar, environmental conditions and extent of pollination. In general, primary berries of flowers ripen first and have maximum seeds. The primary flowers may have little pollen than the later flowers of the same variety. However, stamens are so placed that they crack open and scatter pollen onto pistils. The pistils remain receptive for many days and bees usually get pollen from other flowers to complete the fertilization of primary ones. Pollination of all pistils of a flower is necessary for the maximum berry size.

Fruit growth

At the beginning of the season, average period from flower opening to berry maturity is about 28 to 31 days, whereas in mid-season, it is about 25 days. Generally, the ever-bearing varieties mature in 20 to 25 days in long days but may take 55 to 60 days in high temperature of mid-summer in autumn. Since its fruits show unusual growth pattern, different research workers have given different views on the patterns of its growth. In general, the berry growth in strawberry is attributed to the growth of achenes and receptacle. Initially, the berry growth was reported to be sigmoidal but later some workers reported it to be double sigmoidal. However, the pattern of fruit growth is basically a varietal character and is greatly influenced by climatic conditions. For example, Ozark Beauty exhibits both single and double sigmoidal growth curves under different light intensities. In double sigmoidal growth, the first peak of growth is influenced by the achene development and is associated with initial receptacle growth. The second peak is associated with the transport of carbohydrate assimilated into the receptacle, which rapidly enlarges. On the maturation of embryo, rate of receptacle growth increases and ripening is initiated.

At full maturity, berry acquires a definite shape. The strawberry fruit has eight common berry shapes, viz. oblate, globose, globose-conic, conic, long conic, necked long conic, long wedge and short wedge (Figure 6). The general shape of berry is indicated to some extent by the shape of the receptacle. However, berry shape on the cluster depends on their position and is influenced by the extent of pollination, fertilization and weather at the time of fruit development.

Generally, primary berries tend to be irregular and secondary much uniform in shape.

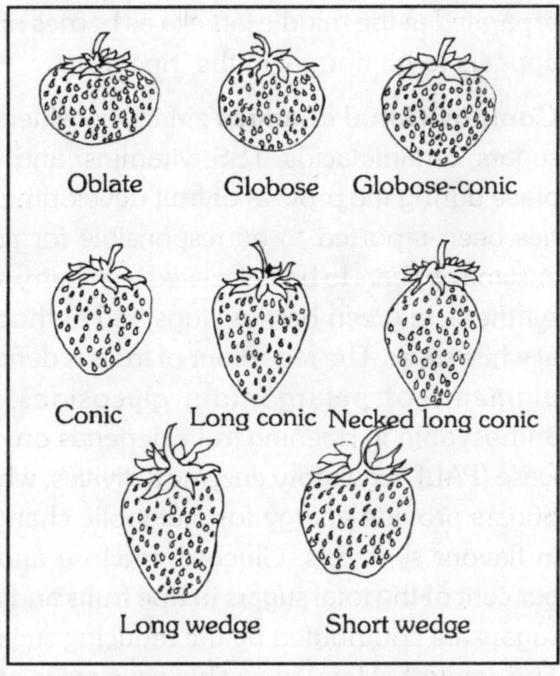

Figure 6. Common shapes of the strawberry fruit

Fruit ripening

Strawberry fruits grow rapidly and may take 20-60 days for ripening, depending upon fruit habit of a cultivar, environmental conditions and pollination.

Strawberry fruit is a typical non-climacteric fruit, yet it undergoes distinct changes during ripening. Its size goes on increasing during ripening process. During ripening process, simultaneous changes in colour, flavour and texture take place. Fruit can acquire attractive colour in storage but changes in texture, flavour, TSS and acidity fail to develop. In general, the following changes occur during the fruit development and ripening process of strawberries.

Cellular changes : The cells of green fruits contain dense cell walls, small vacuoles and starchy grains in plastids. These cells have tubular proliferation of the tonoplast which becomes extensive as the berries ripen. These proliferations may be involved in the synthesis of anthocyanin pigments. At white stage of fruit development, cells expand, plastids degenerate and most of the starch disappears. During ripening, the cells are connected only by small projections at the tips of the cell. In ripe fruit cells, cell walls are swollen. Similarly, parenchymatous cortical cells continue to enlarge and become separated at the middle lamella as berries ripen but the mitochondria appears to be normal in the ripe fruits.

Compositional changes : Many considerable changes in pigments, sugars, organic acids, TSS, vitamins and volatile constituents take place during the process of fruit development. Although, no enzyme has been reported to be responsible for its softening, yet cellulase enzyme appears to be associated with berry softening. The chlorophyll synthesis in green berries stops and anthocyanin begins to develop at white stage. The red colour of fruits is derived from the anthocyanin pigments of pelargonidin glycosides. The accumulation of anthocyanin in ripening fruits depends on phenylalanine ammonia lyase (PAL) regulatory enzyme activities, which is a varietal character. Sugars provide energy for metabolic changes and play pivotal role in flavour synthesis. Glucose, fructose and sucrose account for 90 per cent of the total sugars in ripe fruits and about 80 per cent of total sugars are contributed by the reducing sugars, fructose and glucose. The amount of total sugars between green and ripeness stage doubles (2-5 mg/g fresh-weight basis) during fruit development.

Like sugars, organic acids are important for flavour synthesis in strawberry. The sugar:acid ratio is more often used as an index for consumer acceptability. Similarly, acids also determine fruit pH, contribute to colour stability and inhibit enzyme activity. About 33 per cent organic acids are found in strawberry but citric acid is the primary acid found in all stages of fruit growth and malic acid is found only in traces. The concentration of these acids decreases as the fruits approach maturity. Acidity increases modestly to a maximum in mature green fruits before declining rapidly in later stages of ripening.

The total soluble solids (TSS) of berries increase steadily during development from 5 per cent in small green fruits to 7 to 8 per cent in over ripe ones but these are influenced invariably by cultivar and environmental factors. Polyphenol content (chlorogenic, catechin, coumaric acid) decrease with the advancement in fruit growth, resulting in decrease of about 80 per cent activity of polyphenol oxidase and peroxidase enzymes. The vitamin C content increases as the fruits ripen. About 35 volatile components have been identified to impart flavour in strawberry. However, ethyle hexanoate, ethyl butanoate, methyl butanoate, methyl hexanoate, hexyl acetate, ethyl propionate and 3-hexenyl acetate are the chief compounds imparting much of aroma to ripe fruits, their contents being maximum at ripe stage.

Physiological changes : Though, strawberry fruits do not show a respiratory climacteric during ripening process, yet the respiratory rate of strawberry seems to be quite high. Being non-climacteric fruit, no change in rate of ethylene production even after softening of fruits has been reported.

Factors affecting fruit growth and development

Many factors seem to affect the growth and development process of strawberries. The major ones are :

Soil type : Soil is an important factor for proper growth, fruit set and subsequent fruit development. The plants perform well on well-drained, sandy loam soils and produce healthy fruits of attractive colour, shape and quality. Fruits produced on soils having very low or high pH may be of variable shape, colour and often poor in quality.

Climate : Climate is considered as one of the most potent factors for strawberry cultivation. Of the different climatic factors, light and temperature are more vital for strawberry. Though, present-day cultivars can be grown in different climatic conditions but basically it is temperate fruit. Under cool climate, strawberry grows slowly and takes more time to bear flowers and fruits. In warmer climates, fruits become ready for harvesting within six months after planting but the quality of fruits in comparatively poor. Runners can only be produced in cool climate but not in warmer areas. Similarly, the life of plantations in cooler regions may be 3 to 4 years, whereas only one fruiting can be taken in warmer regions. Sunny days and cool nights favour development of better flavour in fruits than those grown under cloudy, humid days and warmer nights. The shape and size of the fruits are also influenced invariably by edaphic and environmental factors.

Mulching : Mulched plants flower early and produce fruits of better quality. The fruits develop good size and attractive colour and are relatively free from rot under mulch.

Water supply : There should be rigid control on amount and frequency of irrigation. Being shallow-rooted crop, strawberry requires less but more frequent water. Excessive irrigation should be avoided but frequent irrigation during the fruit development is beneficial for proper development of berries. However, proper drainage must be ensured throughout the growing season. Similarily, irrigation should be avoided during ripening of the fruit as it may affect their quality.

Nutritional balance : Proper amount and time of fertilizer application is essential for fruit growth and development. Excess of nutrients especially nitrogen should be avoided as it may result in vigorous growth at the cost of flowering and fruiting.

Provision of pollinators and pollinizers : Though most of strawberry cultivars are self-fruitful, yet they require suitable pollinators and pollinizers for good fruit set and subsequent fruit growth. Provision of a suitable pollinizer and pollinators like honey bees and bumblebees is therefore essential in its plantations for proper pollination. Improper pollination may lead to the development of "off" type (*nubbins*) fruits of different shapes. Thus, unavailability of a pollinizer or pollinator affects fruit development.

Seeds in berry : The number of developing seeds in a berry have strong simulative influence on fruit growth. It is probably due to mobilization of assimilates and auxin from the achenes. The development of receptacle is also directly proportional to the number of fertilized achenes. The partial removal of achenes results in abnormal development of fruits. Thus, minimum number of achenes is required for fruit development and ripening of fruits.

Position of the fruit : The position of fruit in a cluster also affects the size of fruit. Usually, primary fruits are larger, whereas the secondary and tertiary are progressively smaller, which may be due to more number of achenes in primary and secondary berries than the berries developed from tertiary flowers and *vice versa*.

Use of bio-regulators : Many bio-regulators may affect growth and development of fruits but auxin is the dominating hormone regulating growth, development and ripening in strawberry fruits. Developing achenes are good source of auxins (IAA) and number of achenes and their removal exert strong inhibitory influence on the fruit size. Application of synthetic auxins, like NAA, can restore the growth of the receptacle. GA_3 (50 ppm) also acts synergistically with auxins in promoting growth and ripening of fruits.

Chapter 17

Physiological Disorders

Some plant and fruit disorders make strawberry cultivation less remunerative to the growers. Important disorders of strawberry are given below :

Fasciation : The abnormal flattening and enlargement of stems of fruits and fruits and witch's broom appearance of the plants is a major trouble in strawberry. The affected plant produces little marketable fruits and no runners. It results from unfavourable growth conditions in late fall, when days become too short for normal development of a particular variety. In Fasciation, flower bud broadens and in severe cases, no fleshy fruit develops in the spring at all and the plants are barren. The berries sometimes develop into a typical coxcomb shape. Fasciation is a varietal characteristic as some varieties like Missionary and Black more never fascinate. Fasciation may also result from insect damage, disease attack, and lack of pollination, humidity and frost. Proper management of strawberry plantations and provision of suitable pollinator and pollinizer helps in minimising the problem of fasciation.

Fruit malformation : Production of malformed fruits can be commonly seen even in recognized varieties of strawberry, in which common shape of berry is changed. Usually, primary and secondary flowers produce more malformed fruits due to the undeveloped achenes at the distal end of the receptacle than the tertiary and quaternary ones. Many factors are associated with fruit malformation but planting of more vigorous plants, high N levels, insufficient pollination and lack of growth promoting substances are the major ones. Provision of adequate pollinizer and honey bee hives in the commercial plantings may reduce the problem of fruit malformation.

Similarly, planting young and less vigorous plants and avoiding excessive application of N fertilizers also reduce the problem of fruit malformation to a greater extent.

Phyllody : Phyllody is the main abnormality of flowers and fruits of strawberry. It is mainly caused by mycoplasmal infection. In this malady, the flowers and fruits get flattened and fascinated. It first affects the opening of flowers and subsequently the fruit set. The petals of the affected flowers may be thick and develop poor colour. Since, phyllody is difficult to control, it is better to uproot and burn the infected plants.

Albinism : It is the most serious disorder of strawberry fruits, occurring primarily at the time of ripening. It has attained an alarming situation in USA, Belgium and the Netherlands. Fruits suffering from albinism appear bloated and develop white or pink areas on their surface and the pulp remains pale. The fruits have poor flavour and tend to be more acidic. The affected fruits develop normally but do not ripen uniformly and show waxy appearance. These fruits are liable to severe damage during harvesting and become highly susceptible to fruit-rot during storage. Albinism tends to develop in densely growing crops, dense plantings or due to excessive fertilizer application. It occurs frequently in fields during peak fruit production in localities experiencing warm weather followed by overcast skies in the preceding season. It generally develops in crops grown in sandy, low pH soils and soils with high N, K, Ca contents and sometimes P content also. This disorder is more severe in closed tunnels (with little ventilation). Similarly, its incidence is more in plantations, provided with black film mulch than in white films, probably due to high N mineralization under black film, which generates higher temperature. It also seems to be a varietal feature as Elsanta is worst affected by albinism though it appears on Darselect, Eros, Darstella, Gariguette and Dukat also but with lesser intensity.

Selection of a suitable variety, avoiding dense planting and excessive application of fertilizers, particularly N, K etc., are some measures to reduce its menace.

Chapter 18

Important Cultural Practices

Some intercultural practices are important for commercial and profitable cultivation of strawberry. These cultural practices are briefly described below:

Pre-conditioning : Pre-conditioning of strawberry runners is necessary in regions of high latitude and altitude before winter (October-November) planting. It helps the plants to grow vigorously.

Bud and shoot thinning : Sometimes there is excessive formation of buds and shoots, which results in the production of small sized berries. Thus, removal of one or two buds/plant is advocated for proper development of berries. This practice improves fruit yield and quality as well. However, excessive removal of buds may be harmful as it may lead to low yield. Similarly, excessive shoot thinning should also be avoided. Thinning of 2-3 shoots/plant after 13-14 months of planting may increase weight of individual berries.

Removing flower stems : One should remove all flower clusters appearing in the newly-planted runners for getting healthy plants for better flowering and fruiting subsequently. Healthy plants can also tolerate heat and drought and produce more runners. It is more desirable to remove early blossoms in cultivars which are shy in the production of runners. However, flower stems should only be removed once in cultivars producing luxuriant plant growth. While removing blossoms, whole stalks may be picked off, not just the individual flower. The blossoms of fall bearers may be removed 60-80 days after planting. The flower thinning increases the runner production but reduces the yield. Therefore, for fruit purpose, flower thinning in spring should be avoided and if necessary, only tertiary flowers should be removed but not primary and secondary ones.

Flowers are usually removed manually. Manual deblossoming is a laborious and cumbersome practice and hence different chemicals should be used for deblossoming in strawberry. Single application of gibberellic acid (GA_3), 50 ppm, in June and July inhibits flowering, promotes runner development and increases the number of marketable runner plants for the next year. Similarly, application of ethrel (200-300 ppm) at the initiation of flowering also reduces the production of flowers.

Defoliation : It is the practice of removing leaflets of a plant. It is usually done after harvesting, primarily to improve plant health for next fruiting. In this practice, the leaves of first year crop are not removed but defoliation of second and third year crop is necessary. Usually old, yellow and diseased leaves are removed. In general, excessive defoliation (removal of two-thirds foliage) may be avoided as it may reduce the yield and affect the fruit quality.

Thinning of plants : It is virtually necessary to remove diseased, unwanted and thrifty plants. Thinning helps in reducing the crowded condition in the bed and provides sufficient space for the remaining plants that will bear fruits next year.

Removal of runners : Strawberry produces plenty of runners when grown in matted rows because the plants get sufficient space for the production of runners. These runners should be removed immediately before they root. If rooting has already taken place, the runners may be chopped out with hand hoes or removed by sickles. Similarly, in hill culture system, the runner production is less but if they appear, these should be removed immediately. The removal of runners ensures better growth of the mother plants, which subsequently helps in higher yields in the following flowering or fruiting season.

Cleanliness : After harvesting, all weeds, hay, straw or other mulch material should be removed from the fields immediately. The mother plants usually make rapid growth, develop good root system and initiate flowering immediately after harvesting for next season's crop. Hence, all the beds should be fairly cleaned immediately because delaying in cleaning may adversely affect the growth and development of crown, and subsequent flowering and fruiting in the following season.

Renovation of old plantings : Strawberry plantings that are vigorous and reasonably free from weeds can produce two to three successive crops in temperate climate. Therefore, renovation of first planting should be done because the cost of renovation is usually lesser than the cost of a new planting. Renovation involves mowing of plant tops immediately after harvesting, narrowing the rows, plant thinning and application of manures and fertilizers as per soil type, age, vigour, cultivar, region and other factors.

Calyx or hull removal : Removal of calyx from strawberry fruits is known as *capping*. Calyx usually adds beauty to strawberry and intact calyx with fruit is liked by consumers greatly and hence berries are always picked with calyx if they are to be consumed as fresh fruit. However, for processing purpose, the calyx (cap) and stem should be left in the field because it hinders the process of pulp making. Thus, capping of berries is necessary before processing. It can be done by hands or by machines.

Chapter 19

Use of Bio-regulators

Use of plant bio-regulators has become a common practice in strawberry for modifying various physiological processess. Different plant growth substances have been used with advantage in strawberry. A brief account of the substances and their effects have been given hereunder:

Plant growth and runner production : The bio-regulators modify plant growth and runner production in strawberry grown in open fields or under protected structures. Among different growth regulators, GA_3 has been proved as a wonder bio-regulator for enhancing the growth of different cultivars. The amount and time of application of growth regulators depend on the variety, locality and purpose of the grower. In general, spraying of GA_3 (75 ppm) to one-year-old fall bearers promotes vegetative growth and runner production. The increase in plant growth and runner production with GA_3 application may be due to inhibition of flowering. In addition to GA_3, foliar or soil application of PP_{333} (cultar) in strawberry is also beneficial for better growth and increased runner production. Similarly, application of tricontanol (5 ppm) before flower emergence in March also increases leaf area, leaf number and runner production.

Breaking dormancy of runners: The cold-stored strawberry runners usually undergo dormancy and do not establish properly in the field. Treatment of the runners with 0.1% hydrogen cyanamide (HCN) and 3.0% potassium nitrate (KNO_3) breaks the dormancy of runners. Treated plants establish in the field properly.

Flower thinning: Deblossoming is a highly useful practice in strawberry for maximising yield. It is usually done manually, which is a highly laborious and cumbersome practice. To avoid this, one can use different chemicals for flower thinning in strawberries also. A single spray of GA_3 (50 ppm) before flower initiation is the best treatment to suppress flowering. Similarly, spraying of ethephon (500 ppm) at flower-initiation stage is equally effective for smothering of flowering in different cultivars.

Fruit set : Fruit set in strawberry is influenced by many factors, especially by pollinizers and population of insect-pollinators. However, it can be increased by application of auxins, especially by NAA (50 ppm). The NAA should be applied when flower initiation has just started.

Crop regulation: Mulching is the easiest and cheapest practice for promoting flower initiation in strawberry. It helps to improve soil temperature and protect floral primordia from frost damage. However, some growth regulators also promote flowering, subsequent ripening and harvesting of berries. Application of GA_3 (50-75 ppm) in winter is the most effective practice to induce earliness in most of the cultivars.

Induction of seedlessness in berries : The strawberry fruit produces plenty of seeds and these are crunchy in nature. While eating, some consumers feel reluctant to take plenty of seeds inside their stomach. Therefore, production of seedless fruits is sometimes considered desirable. In the past, many attempts were made to induce parthenocarpy in strawberries but the success was limited. However, some research workers were able to produce parthenocarpic fruits using growth regulators. Application of NAA (0.05 and 0.10 mg/ litre) in a lanolin emulsion to emasculated flowers resulted in 100 per cent parthenocarpic fruit development. Though, the application of NAA may induce some abnormalities like reddening, lengthening, twisting, and splitting of tissues in the berries.

Improvement in berry yield and quality : Application of plant growth regulators is one of the best techniques to improve yield and fruit quality in strawberry. A single spray of GA_3 (50-75 ppm) prior to flower-initiation stage is the most effective dose to get higher yield of quality berries.

Chapter 20

Insect-Pests and their Management

Strawberry is attacked by many pests insects which vary widely in their destructiveness and distribution. Many of these are worldwide, while some are more serious in a particular locality. Major pests of strawberry and their management is given below :

Insect-pests : A variety of insects infest strawberry leaves, buds, fruits and roots, thereby causing serious damage. Some of those infesting strawberries are :

Red spider mite (*Tetranychus urticae*) **:** It is the most serious pest of strawberry. The mites are brick red in colour and usually have two darker spots on the back. It over-winters in adult stage in soil and plant debris and becomes active during April. It lays white eggs underneath the leaves. The eggs hatch within 4 to 5 days. The damage is caused both by nymphs and adults by sucking the sap from the leaves, causing a rusty-brown colour to them. The plants infested by Red spider mite are stunted and yield is reduced. Spraying of chloropyriphos and aldicarb is very effective for controlling Red spider mite. Aldicarb (10G) @ 5 kg/ha is quite effective in controlling Red spider mites. Control of these mites through the biological means has also been found to be very effective. Some useful predatory mites for the control of Red spider mite are, *Amblyseius fallacis*, *Phytoseiulus persimilis, Feltiella acarivora* and *Typhlodromus pyri*.

Mite (*Phytonemus pallidus*) : It is a troublesome pest in warm areas, which feed on young folded leaves. The damage is caused by sucking of sap from leaves, resulting in rough rust coloured patches and leaf edge curling. The older leaves become puckered and wrinkled. Later, the crown of the plant may be infested, turning pale-yellow and ultimately dying. A single spray of endosulfan (0.5 per cent) is quite effective to control it.

Blossom weevil (*Anthonomus rabi*) : In some localities, blossom weevil is also called elephant weevil. These are greyish-black in colour and lay eggs in unopened flower buds. The damage is caused by making holes in the flowering stalks and buds, causing whole truss to wilt and die. Spraying of systemic pyrethroids like deltamethrin (0.05 per cent) and sumicidin (0.025 per cent) just before flowering saves the plants from its damage.

Root weevil (*Otiorhynchus rugostriatus*): This is a wingless weevil, whose grubs are more harmful than the adults. The small white legless grubs feed on small rootlets in winter and spring and make deep tunnels at the base of the crown as a result of which the plants collapse completely. The adults feed on foliage at night but the damage is less severe than that caused by the grubs. Application of carbofuran (6-10 kg/ha) or parathion (0.015 per cent) around the plants, reduces the pest population and damage caused by the weevils. The pest population can also be controlled by a nematode, *Heterorhabditis heliothidis*.

Leaf rollers : Many leaf rollers infest strawberries, but *Planototrix exessana* causes the maximum damage. It is small reddish-brown moth, which deposits small eggs on leaves. The caterpillars, on hatching, cause the leaflets to fold at the midrib and feed within this shelter. Only the epidermis is eaten but continuous feeding causes the entire leaf to turn brown and die. Leaf roller can be controlled by early season spraying (August-September) with endosulfan (0.05 per cent) followed by fortnightly spray of dichlorvos (0.04 per cent) from November to February.

White grubs : White grubs also cause severe damage to strawberries. The larvae are dirty white with brown heads, usually coiled and with distinct legs. They eat off roots and kill plants, usually between planting time and runner formation. Drenching of soil with chloropyriphos

(0.02 per cent) reduces the damage of white grubs. Geese, chickens or hogs may also destroy the grubs.

Thrips (*Thrips atratus)* : Nymphs and adults cause damage by sucking sap from growing points, leaves and developing fruits. The damage is more severe in late-season strawberries. Severe infestations may result in fruit distortion, leading to down grading and financial loss to the farmers. Spraying synthetic pyrethroids like deltamethrin (0.2 per cent) and cypermethrin (0.5 per cent) frequently reduce the incidence of Thrips and thereby distorted fruits.

Cut-worms : Several species of cut-worms infest strawberries. Large stout stripped mottled or grey caterpillars eat on fruits and make large holes. They usually feed at night and hide during the day. They cut off the young plants at the ground level, and may eat leaves and berries of the established plants. Drenching soil with Chloropyriphos (0.01 per cent) during March is effective for the control of cut-worms. During land preparation, 4 to 5 deep ploughings should be done and the fields dusted with insecticides.

Tarnished plant bug : The adult bug is coppery-brown with piercing and sucking mouth parts. It causes damage by puncturing the berries, causing uneven ripening or "catfacing" of the fruits. It also feeds on individual achenes (seeds) and destroys their contents. Spraying of endosulfan (0.02 per cent) is effective for its control.

Leafhoppers (*Aphrodes bicincta*) : The leafhoppers cause damage by sucking sap from growing leaves and flower buds. The affected leaves turn pale-yellow and curl inwards from the margins. The affected leaves also develop brown colour and often have dead tips. Leafhoppers mainly act as vectors for transmitting viral diseases. Predators like syrphid flies and green lacewings keep the population of leafhoppers under control.

Other pests

Some nematodes, snails and slugs and birds also cause severe losses to strawberries. They are:

Nematodes : Nematodes are thread-like worms, too small to be seen easily with naked eye. These are usually more destructive in sandy soils than in clay or soil with a high organic-matter content. Some species live in soil and several attack roots of strawberry. Many species live on the parts of plant above groun, causing serious damage to buds and leaves. Sting, root-knot, bud and lesion nematodes cause considerable damage to strawberry plantations.

Sting nematode (*Belonolamus* sp.) is more destructive to strawberry than the other nematodes. It occurs mainly in the light soil. It feeds on the surface of roots and are usually dislodged when plants are dug. Root-knot nematode (*Meloidegyne* sp.) penetrates and feeds in small roots, causing knot like enlargements in them. The plants are weakened, producing few runners and fruits. These nematodes enter the roots and can stay alive even when the plants are dug, stored or shipped. They infest all strawberry cultivars in North India. Lesion nematode (*Pratylenchus* sp.) also attacks the roots and makes dark brown small spots initially. Later, the roots may decay, showing damage similar to the one caused by root-rot diseases. The leaf stalk length of the affected plants is shortened and leaves become yellow. The plant growth is reduced remarkably. This nematode may increase the chances of *Verticillium* wilt. Bud nematode (*Aphalenchoides* sp.) lives in most succulent tissues of the buds and in very small leaves. The flower buds are often killed by feeding, resulting in flowerless plants. It causes the bud disease called 'spring crimp' or 'red plant'. Stem nematode (*Ditylenchus dipsaci*) lives in plant stems, causing damage by sucking sap from them. It results in shortening and thickening of stems and leaf stalks.

All soil-inhabiting nematodes can be checked by fumigating the soil with nematicides like methyl bromide or by the application of microgranular nematicides like thimet, temik etc. Similarly, root nematodes can be controlled by Basamid (dazomet) application and by solar heat treatment. Nematodes in roots, leaves and buds can be killed by hot water dip of dormant plants at 61° C for 2 to 3 minutes. Spraying of simazine (2 or 3 kg/ha), a herbicide, can also control the stem and bud nematodes. Similarly, population of nematodes is reduced if mustard, rye or maize is used as green manure in strawberry.

Snails and slugs : These are elongated, slimy mollusks with or without a shell. They eat out holes in foliage and berries, leaving slimy trails. They are highly troublesome during wet seasons and cause maximum damage at night. The best way to control the snails is to use common salt sprays. Beer pit falls and metaldehyde baits are also quite effective in control of the snails and slugs.

Birds : Some birds, especially sparrows, peacocks and parrots cause heavy losses to strawberry at the time of berry ripening. The whole berry is picked and eaten up by these birds. The damage by birds is more alarming than that caused by insect-pests or diseases. Use of bird scarer and tying of reflecting ribbons in fields can reduce the damage caused by them. Similarly, covering the entire field with nylon netting is very useful for avoiding the damage caused by birds (Plate 14). The birds are very clever and soon become accustomed to the measure being used for scaring them and hence no scaring device should be used continuously for a long time. Some cultivars are resistant to some insect-pests and nematodes and their cultivation must be encouraged. A list of some strawberry cultivars is given in Table 7.

Table 7 : Strawberry cultivars resistant to some insect-pests and nematodes

Insect-pest/nematode	Cultivar
1. Strawberry aphid	De Norte, Yaquina, Benton
2. Weevil	Totem, Cheam, clones of *Fragaria chiloensis* etc.
3. Two spotted spider mite	Marshall, Aliso, Apollo, Lassen, Cardinal, Comet, Blakemore, Siletz, Stelemaster, Muir etc.
4. Root-knot nematodes	Frenso and Torrey

Plate 13. Strawberry flower

Plate 14. A view of protecting strawberries from bird damage by installing nylon netting over the field

Chapter 21

Diseases & their Management

The problem of diseases is one of the most limiting factors in the profitable cultivation of strawberry. Many leaf, root, fruit and viral diseases have been reported to cause heavy losses in strawberry. Important diseases have been birefly described below :

A. Foliage diseases

Leaf-spot : Among various foliage diseases, damage to leaves is probably the greatest by the leaf-spot. A fungus, *Mycosphaerella fragariae*, causes it. In leaf-spot, the spots are small and round with a white or grey centre and purplish margins. In severe infestation, it may cause stunting of plants and occasional fruit infection. Its incidence is severe during rainy season, when relative humidity is usually high. "Black seed" (1-10 black spots around a group of seeds) may occur in fruits. Following control measures have been suggested to control leaf spot :

i) Follow a clean cultivation,

ii) Grow resistant varieties like Fairfax, Aroma, Redstar, Temple, and Surcrop etc.

iii) Collect and burn the infected leaves to avoid the further spread of the disease.

iv) Spray Bordeaux mixture (1%) or Dithane M-45 or Difolatan at fortnightly interval, 2-3 times in a season

v) Give a single spray of carbendazim (0.05%) between the spray of Dithane M-45 or Difolatan.

Leaf-scorch : It is another serious disease of strawberry foliage. It is caused by a fungus *Diplocarpon earliana.* It is a common disease in midhill areas, during April-May. Numerous irregular, purplish spots with white centres, surrounded by halo, appear initially on the lower surface of the leaves. These spots may enlarge, coalesce, and cover the entire leaf. The spots may also appear on the petioles and runners as well. It is difficult to control it with spraying and dusting because spots appear on lower surface of the leaves. Howard 17, Surecrop, Rockskill, Aroma, Empire, Catskill and Sunrise are resistant cultivars against leaf-scorch and these should be encouraged in localities having the problem of leaf-spot disease.

Powdery mildew: It may affect strawberry during dry periods. A fungus, known as *Sphaerotheca macularis,* causes it. Due to its infection, the leaves show a characteristic upward curling. As the disease progresses, the underside of leaves reddens and affected parts may be killed and the flowers and green berries may not develop properly. The fungus also infects flower stalks, flowers and developing fruits, covering them with white powdery mass, which may lead to wilting and drying of flowers and fruits. Fortnightly spraying of bayleton (0.05 per cent) and Nimrod (0.1 per cent) between May and September are quite effective in its control. Sulphur and copper fungicides are also equally effective. Catskill, Dunlop, Klondike, Marshall and Sparkle are some varieties resistant to powdery mildew.

Leaf blight : Among various foliage diseases, leaf blight is known to cause least damage to strawberry. It is a fungal disease, caused by *Dendrophoma obscurans*. In leaf blight, the spots are initially purple coloured but later develop into a dark brown areas, surrounded by purple rings. Often V-shaped infection occurs with wider part at the leaf margins. It may cause stem-end rot in warm weather. Earlidawn and Howard 17 are resistant varieties.

Angular leaf spot: It is a bacterial disease, caused by a gram-negative bacterium, *Xanthomonas fragariae*. It is a serious disease in humid weather. The planting in early autumn and late spring suffers worst from it. The leaves turn yellow with angular spots of variable intensity. Severely infested plants show mucoid xanthomonad type growth and die in a short time. Spraying of Bordeaux mixture (1 per cent) may keep it under control.

B. Root diseases

Red stele : It is the most destructive disease of strawberry. It causes heavy loss in cool and moist soils having poor drainage. It is also known as red core or brown core. A fungus, *Phytophthora fragariae*, causes it. The fungus enters the central parts of the roots, the stele, which turn reddish, while the cortex of roots appear normal. In dry season, infected plants may die before blossoming but in wet season, they may blossom, but die before the ripening of fruits. The affected plants at the first show dull, bluish-green colour, wilt soon and may die in a few days. In well-drained, sandy soils, red stele is never a serious disease. It is spread by drainage water, machines and tools, and by infected plants. The fungus causing red stele has at least five races and they differ in their ability to infect different cultivars. Therefore, a cultivar resistant in a particular locality may be susceptible in the other. Some cultivars resistant to one or the other races of red stele are Temple, Fairchild, Red glow, Climax, Cambridge Vigour, Marshall and Little Scarlet.

Improvement in drainage system, ridge planting, crop rotation, covering the plants with cloches and use of resistant varieties are important cultural practices that reduce red stele incidence. However, soil drenching with disinfectant like Ridomil and treating runners with Bordeaux mixture (1 per cent) or copper oxychloride (0.1 per cent) for 15 to 20 minutes before planting also reduces the chances of occurrence of red stele in field.

Verticillium wilt : It is a soilborne vascular disease, caused by a fungus, *Verticillium albo-atrum*, which is more active in cool weather. The fungus invades roots and interferes with the movement of water to the leaves. Internal browning of vascular tissues at the base of the crown is its major symptom. The outer older tissues wilt and dryup at margins and become reddish to dark brown. Only a few leaves develop, which tend to be stunted and may curl up along the mid vein. The central young leaves are small, new roots from the crown are short with blackish tips and plants look dry and may suddenly collapse. The plants are often stunted and flattened as if suffering from lack of water. Many crop-growing soils contain one or more races of Verticillium wilt. The fungus causing wilt infects several host plants, therefore strawberry should not be planted in fields where solanaceous crops have been grown previously. Once the fungus is established in an area, it may remain alive for 20 to 25 years. However, soil fumigation is an ideal way to keep the disease under control. Similarly, crop rotation and planting of resistant varieties like Howard 17, Catskill, Marshall, Senga Sengana etc., are other methods to get rid of wilt.

Black root-rot : It is a name given for several problems that produce similar symptoms like stunted plants, wilting and dying of plants when berries are half grown. The infected plants develop smaller roots. The main root may have dark spots or patches and the tip of main feeder root is dead. The plants lack vigour and produce few runners. It may be due to nematodes, soil fungi, winter injury, heaving, fertilizer burn and too much water, salt or alkalinity in the soil. Planting of healthy runners, following crop rotation and use of soyabean as a green manure etc., may save strawberry plants from black root-rot infection.

6. Fruit diseases

Grey mould or botrytis rot : It is a serious disease of strawberry both in the field and in the storage. It is caused by a fungus called *Botrytis cinerea.* The disease appears as light brown, soft spots on green and ripening fruits. The berries dry out, become tough and are covered by a dusty fungal growth (Plate 15). The infection is common on raw and ripe fruits, resulting in complete rotting of the fruits. About 50 per cent of fruit loss in strawberry may be due to botrytis rot. A single infected fruit may spread the rot to all the fruits during transit or storage. The following remedial measures should be followed for its control:

1. Do not allow the fruits to touch the soil.
2. Provide suitable mulch at the time of fruit ripening.
3. Avoid excessive use of nitrogenous fertilizers.
4. Avoid excessive irrigation and contact of berries with water.
5. Ensure proper drainage.
6. Proper sanitary measures should be followed.
7. Give a spray of carbendazim (0.05 per cent) just after the opening of flowers.
8. Remove all the disease-infected fruits before storage or long distance transportation.

Leather rot : It is also a fungal disease, caused by *Phytophthora cactorum*. It is a serious problem in areas having high humidity and soil with poor drainage. The berries on or near the soil turn brown and appear water-soaked. Berries at all stages of maturity become hard and leathery. Proper drainage and use of broad-spectrum fungicides like benlate, captan, or thiram etc., can save the fruits from leather rot.

D. Viral diseases

Several viral diseases cause damage to strawberry plants. They may produce mottling, mild yellow edge, crinkle, and vein chlorosis, leaf stunting or dwarfing. Two general types of viruses affecting strawberries are killer and latent viruses. The killer virus produce severe symptoms and kills individual mother plants and attached runners. The latent virus is more often responsible for progressive weakening of a planting stock. A latent virus by itself may produce no obvious symptoms but combination of two or more viruses may produce yellow leaves, crinkled foliage or severe stunting of plants. Viral diseases are transmitted by insect vectors, chiefly by aphids. However, leaf-hoppers and nematodes also transmit some viruses. The main viral diseases of strawberry are yellow edge, crinkle, arabis mosaic, green petal and bronze leaf wilt. Out of these, yellow edge and crinkle are transmitted by aphids, green petal and bronze leaf wilt by leaf-hoppers and mosaic by nematodes.

Viral diseases may be controlled easily if insect vectors are kept under control. Further, clean cultivation and crop rotation also help to eliminate the chances of viral infection in strawberry plantations. Similarly, virus free planting material should be used for establishing strawberry plantations and if any infected plant is noticed in field, it should be uprooted immediately and burnt to avoid further spread of viral diseases.

The control of various diseases by fungicides is really a tough job. It involves a large amount of money and technical expertise. Further, the residues left on plants and fruits are very harmful to man. Therefore, use of resistant varieties should be encouraged. A list of some cultivars resistant to various diseases is given in Table 8. These cultivars can be effectively used as one of the parents in hybridization for incorporating the source of resistance against a particular disease.

Table 8 : Strawberry cultivars resistant to some diseases and physiological disorders

Sl. No.	Disease	Resistant cultivar
1.	Ramularia leaf spot	Aroma, Howard 17, Fairfax, Midland, Dabreak, Blakemore, Lateglow, Surecrop, Redstar, Temple, Albritton, Headliner, Lassen, Valentine, Empire, Elista Cyclone, Hood, Tangi, Missionary etc.
2.	Leaf scorch	Howard 17, Dorsett, Rockhill, Fairfax, Empire, Catskill, Dixieland, Fletcher, Geneva, Trumpter, Sunrise, Midland, Blakemore, Surecrop, Redstar, Temple, Albritton, Headliner, Lassen, Valentine, Empire etc.
3.	Leaf blight	Earlidawn and Howard 17
4.	Powdery mildew	Catskill, Surecrop, Lassen, Valentine, Tioga Brighton, Columbia, Umpire, Sparkle, Florida Ninety, Marshall, Orland, Gorella, Cambridge Favourite etc.

Sl. No.	Disease	Resistant cultivar
5.	Verticillium wilt	Vemilon, Climax, Temple, Allstar, Delite, Lestar Guardian, Hecker, Hood, Siletz, Catskill, Sunrise, Robinson Howard 17, Marshall, Gem, Senga, Sengana, Jupsa, Sierra, Lateglow, Tribute, Tristar, Totem, Midway, Earliglow etc.
6.	Red stele root rot	Sparkle, Sunrise, Midway, Siletz, Del Norte, Juspa, Sunrise, Climax, Redgauntlet, Little Scarlet, Marshall, Gorella, Lavo, Templer, Redchief, Surecrop, Crusader etc.
7.	Viruses	Northwest, Howard 17, Fairland, Blakemore, Teenase Beauty, Siletz, Klondike, Missionary, Robinson, Shasta, Tioga, Temple, Pugget Beauty, Columbia, Cheam, Totem, Delite, Shuksam Earlidawn, Raritan, Redstar, Tristar, Gem, etc.
8.	Botrytis rot	Earliglow and Tyee
9.	Black rot	Siletz
10.	Fruit rots	Solana, Cascade, Columbia, Mollala, Jucumba, Juspa, Valentine, Senga Precosa etc.
11.	Variegation	Marshall, Klondike, Lassen, Climax, Midland, Vermilion, Suwannee etc.
12.	Fasciation	Blakemore, Missionary and Klondike etc.

Chapter 22
Harvesting and Yield

Harvesting

Harvesting is a very important operation in strawberry. Harvesting time differs with the locality, variety and season of the crop and market facilities. The crop is ready for harvesting in March in sub-tropical plains. But, in cool climate, it matures by mid-June. Its fruits are usually harvested when two-thirds to three-fourths portion of the fruit skin has developed colour. For distant markets, the berries should be harvested while they are still green or white and are hard. However, a little delay in picking may increase the percentage of over rotted berries.

The berries should be harvested in shallow containers because these are highly perishable and damage early if deep and bulky containers are used. Usually, berries should be picked up daily in warm weather and 2 to 3 times a week in cool weather conditions. Berries do not improve in quality after picking. Marketing of insufficiently coloured berries is not conducive. Therefore, berries which are greenish-white at harvesting time, may be exposed to a temperature of 40° C for 42 hours for proper colour development. In general, berries picked at 50 per cent colour development may be slightly smaller, less sweet and with low flavour. For fresh market, leave the caps and 3/8 to 1/2 inch of stem on the berry when picking. Pinch the stem close to the cap with thumb and forefinger. For processing purpose, hold the stem and cap (sepals) with one hand and pull off the berries with a slight twist of the other hand. The berries should be picked in early and in late evening hours of the day. Keep the picked berries in a shade and dust-free place. A shelter bed, even a structure with simply a roof and a wall on one side of the prevailing wind is advantageous in reducing the farm heat of the berries and other adverse effects. The length of time for which the berries can be kept after picking depends on cultivar, degree of ripeness, care in handling and temperature when they are picked and kept. For each rise of 1° C, the life of berries is decreased by 50 per cent, i.e. if berries are kept for eight days at 3° C, their life will only be one day at 30° C. Further, the life of berries after harvesting is short owing to a relatively high rate of respiration. Pre-cooling of strawberries at 3° C for two hours is necessary. It becomes necessary during warm weather when fruits are picked during the heat of the day at a temperature of 30° C or above.

Following points should be kept in mind while harvesting the strawberries :

- Berries may be picked every day in warmer climate and twice a week in a cooler climate.
- Pick the fruits everyday for fresh markets and thrice a week for distant markets.
- Pick berries in the early morning or in late evening hours.
- Pick berries along with calyx for fresh markets and capped berries for processing purpose by pinching and twisting the stem and not by pulling.
- For fresh markets, pick berries which have developed 75 per cent skin colour and for distant markets mature greenish white berries may be picked.
- After harvesting, place the fruits under a shelter shed.
- Pre-cool the berries if harvested during the heat of the day.

The harvesting is done manually in most of the countries. The number of pickers needed for harvesting varies with their efficiency, crop load, weather and the method of harvesting. In general, 5 to 6 pickers are sufficient for picking strawberries from one hectare field. In advanced countries, harvesting is done mechanically. For mechanical harvesting, 'Stripper', a full row self-propelled unit, and the "Clipper", a pull type harvesters, have been developed in the USA .

Pick-your-own berries (PYO) : In most of the advanced countries, labour is always in short supply. Thus, shortage of labour and competition from distant areas have forced many growers to adopt the pick-your-own method of harvesting. The customers drive to the planting sites and they supply their own containers, when sale is made by weight. In some cases, the baskets may be obtained from the growers and used as a unit of sale. Saving in costs of picking, packaging and delivery to the market has greatly attracted the growers to use PYO method of harvesting in the recent years.

Yield : It is very difficult to predict the yield of strawberry because yield is not a constant factor as it is influenced by soil and climatic conditions, variety, planting system, planting density, number of crowns per plant, number of fruits per plant, fruit size and various management practices. The average yield of strawberry in California, Italy, Poland and India was recorded as 30, 12, 6 to 7 and 5 to 7 tonnes/ha, respectively. Tioga cultivar can give an average yield of 5.7 tonnes/ha, Blakemore 5.2 tonnes/ha and Shasta 5.1 tonnes/ha in India. The yield and quality of most of the cultivars in temperate zone is better as compared with the subtropical zones, especially in plains of north India. However, in hills, the yield of better quality is obtained. The judicious use of fertilizers, timely irrigation, proper mulching, control of insect-pests and diseases, use of growth regulators, provision of honey bee colonies are some reliable inputs which may help in boosting the productivity of strawberry in India.

Pick-your-own berries (PYO): In most of the advanced countries, labour is always in short supply. Thus, shortage of labour and competition from distant areas have forced many growers to adopt the pick-your-own method of harvesting. The customers drive to the planting sites and they supply their own containers, when sale is made by weight. In some cases, the baskets may be obtained from the growers and used as a unit of sale. Saving in costs of picking, packaging and delivery to the market has greatly attracted the growers to use PYO method of harvesting in the recent years.

Yield : It is very difficult to predict the yield of strawberry because yield is not a constant factor. It is influenced by soil and climatic conditions, variety, planting system, planting density, number of crowns per plant, number of fruits per plant, fruit size and various management practices. The average yield of strawberry in California, Italy, Poland and India was recorded as 50, 12, 6 to 7 and 5 to 7 tonnes/ha, respectively. Flora cultivar can give an average yield of 5.7 tonnes/ha, Blakemore 5.2 tonnes/ha and Shasta 3.1 tonnes/ha in India. The yield and quality of most of the cultivars in temperate zone is better as compared with the subtropical zones, especially in plains of north India. However, in hills, the yield of better quality is obtained. The judicious use of fertilizers, timely irrigation, proper mulching, control of insect-pests and diseases, use of growth regulators, provision of honey bee colonies are some reliable inputs which may help in boosting the productivity of strawberry in India.

Chapter 23
Post Harvest Management

Grading and Packing : To fetch good price of the produce, berries should be graded according to their size, shape and colour. The fruits of uniform size and colour should be selected for A grade packs, whereas slightly damaged fruits may be packed and sold out as B grade. Before grading, remove all cull, diseased or rotten fruits. In India, sorting is usually done manually. In some advanced countries, pan method of sorting is followed for grading strawberries. The useful grading pan is 25 cm long, 2.5 cm wide and 1.4 cm deep. For grading, the commercial growers run the fruits over a belt, grade it and then jumble back into cardboard trays of different sizes, which can conveniently be stacked for pre-cooling and marketing. Standardized grades promote honest and fair dealing and discourage careless and unscrupulous packing.

The strawberries may be packed in small baskets, tins, polyethylene over-wrapped baskets and crates. Now-a-days, special packing cases of plastics, commonly called plastic punnets have been developed for packing strawberries (Plate 16). About 200-250 g fruits can be safely packed in one punnet (Plate 17 and 18). These packs provide proper ventilation for strawberries because they have many perforations (10-12 holes/punnet). After packing, these punnets are placed in corrugated fibre trays. The punnet filled trays should be kept under shade or shelter to reduce the heat and water loss from the berries. The filled baskets should neither be slack nor so full that berries are crushed by covers or dividers. Packs should be full enough to look attractive and maintain a well-filled appearance for the consumer. Similarly, berries of one cultivar should be packed in one container and should not be mixed with the other. The pack should

Plate 15. Botrytis rot infected fruits of Chandler strawberry

Plate 16. Plastic punnet used now-a-days for packing, strawberries

Plate 17. Strawberry fruits kept in punnet after harvesting

Plate 18. Packed fruits of strawberry

be properly labelled with the name of variety, garden, date of packing etc.

Pre-cooling : The strawberries should be pre-cooled at 3° C or so within two hours of harvesting and after pre-cooling, these should be kept at this temperature for some time. This is especially necessary during warm weather when fruit is picked during the heat of the day and fruit temperature is 30° C or more. This treatment keeps the berries fresh for a longer period and helps to fetch better price in the market. At higher temperature, the rate of respiration goes high, which results in further reduction in the life of berries. Therefore, pre-cooling of berries is absolutely necessary.

Storage : Strawberry is one of the most perishable fruit and therefore, it should be marketed soon after the harvesting. Due to its perishable nature and poor transport facilities, the fruits get crushed during long distance transportation. Therefore, refrigerated transportation and short-period storage of strawberries can give good money to the growers. Further, strawberry fruits are available only for a short time (15 to 20 days) and their proper storage can extend the availability of strawberries for a longer time.

Pre-harvest application of fungicides has been tried to increase the post-harvest life of strawberries but with limited success . Therefore, it is better to store strawberries in controlled atmospheric storage conditions. The temperature and concentration of CO_2 and O_2 in the controlled atmospheric conditions vary with cultivar, but in general, the strawberries can be stored for 10 to 12 days in good condition in controlled atmosphere (CA) storage system at 2 per cent O_2 and 5 per cent CO_2 with temperature at 3° C. Long-term storage of strawberries is possible at 1 per cent or lower concentration of O_2 but at higher concentration of CO_2 (10 per cent or more) it can be stored for a week only. The post-harvest life of berries in storage can further be increased if the plants are sprayed with bavistin (0.1 per cent) during flowering and then storing the berries at 20 per cent CO_2 and 0-3 per cent O_2 concentrations.

Pre-harvest management factors for increasing post-harvest life

Many pre-harvest factors have great effect on the post-harvest shelf life of berries. Thus, management of these factors may increase their post-harvest life invariably. Some of these factors are described below:

Edaphic conditions : To get quality berries, it is advisable to grow strawberries on sandy loam soil, which is porous and rich in humus. Strawberries grown on heavy or calcareous soils are poorly developed and can be infested with diseases. Similarly, pH of soil should be between 5.5 to 6.5 for getting quality berries. Ensure proper drainage in the selected soil. All these conditions of soil enhance post-harvest life of the berries.

Choose ideal cultivar : Grow varieties as per the need of the market. For example, if market is near, grow varieties which cap easily. For processing purpose, varieties which do not cap easily may be grown. For distant markets, varieties with good shelf life and which are resistant to botrytis rot should be preferred.

Proper mulching: Mulching is the most wanted agricultural practice in strawberry. It is particularly required at the time of berry ripening. In addition to moisture conservation and temperature regulation, it avoids the contact of berries with soil and thus minimizes the chances of fruit-rot, which adversely affects post-harvest life of the berries. Use of proper mulch is, thus, equally beneficial for increasing shelf life of berries after their harvesting.

Irrigation : The amount and frequency of water in strawberry plantations should be rigidly controlled. Accumulation of water near the plants is very dangerous to strawberries as it invites fruit-rot, the most serious disease. Thus, proper drainage should be ensured. For this, drip irrigation system is very efficient and may be installed for getting better yield of quality strawberries.

Vigilant check on botrytis rot : Botrytis rot is the most devastating disease of strawberries, occurring in the field and storage. This disease must be controlled effectively in the field itself. If the harvested fruits are devoid of rot, the rotting in storage or transportation will be less, thus increasing the shelf life of berries considerably. For controlling botrytis rot, maintain proper drainage, avoid excessive use of N and water application, and make frequent sprays of carbendazim (0.05 per cent) just after the opening of flowers.

Calcium sprays : Pre-harvest application of calcium is useful to increase post-harvest life of strawberries. Foliar application of calcium chloride (1 per cent) a few days before harvesting enhances post-harvest life of strawberries by delaying the incidence of grey-mould.

Harvesting at appropriate time : To achieve maximum benefits, picking of berries should be done at appropriate date and time. Avoid harvesting of over-ripe berries. Delay in harvesting may result in production of over-ripe or diseased berries, reducing their post-harvest life. For a nearby market, harvesting of berries should be done when three-fourths fruits have attained red colour. However, for distant markets, fruits attaining 50 per cent of colour may be harvested. Similarly, harvesting should be preferred in the morning and evening hours and should be avoided in the hot hours of the day.

Shade : After harvesting, berries should be shifted to shelter shed to remove their heat. It also helps in enhancing the post-harvest life of the berries.

Sorting : All the diseased or damaged berries should be sorted out before grading and packing them. Strawberry is highly perishable fruit and even if a slightly damaged or rotten fruit is packed in a healthy container, other berries are damaged by the next morning. Thus, sorting, grading and packing should be done under close supervision.

Pre-cooling : The berries, after harvesting or packing, may be cooled at 3° C for two hours and may be kept at this temperature before sending to market or storage.

Chapter 24
Value Added Products

In addition to table and dessert purposes, various value-added products can be prepared from the strawberry fruits. Its berries can be used for canning, candy, jam and jelly making and for flavouring ice-creams. Brief procedures for making different strawberry products are given below:

Strawberry preserve and candy : A whole fruit when impregnated with cane sugar or glucose syrup and subsequently drained free and dried is called candy or preserve. For making strawberry preserve, take firm ripe berries of good colour and flavour. Sort them carefully and remove their calyx and stems. Wash berries carefully and place them in a pan containing sugar and some water. Warm the mixture to dissolve the sugar, skimming off impurities in the process. Heat the syrup to a temperature of about 104° C and then cool it to 94° C and then transfer the strawberries into it. To prevent floating of fruit, place wire mesh tray over it and set aside for 24 hours. Now boil the mass again. The preserve is now ready. The freshly made preserves are wholesome and have an attractive appearance. When they are stored for a long period, their natural colour and flavour deteriorate on account of oxidative changes. These should, therefore, be made only during the season unless there are adequate facilities to keep the fruits. Preserves made from frozen fruits are generally superior in colour and flavour to those made from fresh fruits stored at ordinary room temperature.

Canned strawberry : In some advanced countries, canned strawberry is quite popular. For canning, select healthy, attractive and uniform-sized fruits. Remove their stems and wash them properly. Place these fruits in sugar syrup of 50° Brix. Boil these fruits along with sugar syrup slowly. Keep it overnight and boil it quickly before packing. Exhaust can at 82-100° C for 6 to 10 minutes or until temperature in the centre of the can reaches about 79° C. Both tin and glass containers are used in canning industry but tin containers are preferred. Among different containers, Lacquered AR can should be used for its canning, because strawberry fruit has high acid contents. The cans should be sealed properly, processed for 10 to 15 minutes at 115-121° C and then cooled rapidly to 39° C to stop the cooking process and to prevent stack burning.

Strawberry jam : Strawberry jam is popular all over the world. It is made by boiling the fruit pulp with sufficient sugar to a reasonable thick consistency, firm enough to hold fruit tissues in position. As per FPO specifications, jam contains 68.5 per cent soluble solids, 0.5-0.6 acids and SO_2 content not more than 40 ppm. For jam making, select ripe firm fruits, sort them carefully, remove caps and wash them thoroughly to remove the adhering dust and dirt. Crush them between rollers for pulping. Add some water to pulp if necessary. Add sugar to the pulp and boil it along with water and sugar with continuous stirring. Add citric acid to the boiling pulp. Judge the end point by further cooking up to 105° C or 68-70 per cent TSS or by sheet test. Fill it hot into sterilized bottles, cool the bottle, wax and cap them properly. For 1 kg pulp, 0.75 kg sugar, 2 g citric acid and 100 ml of water are required to make good quality jam. The SO_2 (40 ppm) or benzoic acid (200 ppm) can be used as preservatives.

Strawberry jelly : A jelly is a semi-solid product prepared by boiling a clear, strained solution of pectin, containing fruit extract, free from pulp, after the addition of sugar and acid. A perfect jelly should be transparent, well set, but not too stiff and should have the original flavour of the fruit. In jelly making, pectin is the most essential constituent. Strawberry contains pectin but has been classified as a fruit with low pectin and acid and, therefore, commercial pectin or acids like citric, tartaric or malic acid may be added artificially. However, these commercial products have deleterious effect on the final flavour of the product. In general, 0.5-1.0 per cent pectin in the extract is sufficient to produce good jelly. As per FPO specifications, a jelly should have at least 65 per cent soluble solids and 45 per cent fruit with maximum SO_2 of 40 ppm. The strawberry fruits for jelly making should be sufficiently ripe but not over-ripe and should have good flavour. Slightly under-ripe fruits yield more pectin than the over-ripe fruits. Wash the fruits with ordinary tap water to remove dirt and dust. De-cap the berries and extract the pectin by crushing the fruits and boil it for 5 to 10 minutes and add water if necessary. Now, add required quantity of sugar and acid during boiling. The pectin extract should be strained and clarified. Fill it in bottles when it is still hot and then cool to room temperature. Wax the jars and cap them properly.

Other products

Strawberry juice is also made from strawberry fruits. It is considered as a refreshing and soothing drink in summer. Some developing countries also make wine and freeze products from the strawberries. In India, strawberry is consumed as a fresh fruit. However, it is also used for flavouring ice-cream. One and all like the mother dairy ice-cream of strawberry flavour in our country. Now-a-days strawberry flavoured kulfies are also coming in the markets.

Chapter 25

Breeding and Crop Improvement

The increasing trend in the cultivable area under strawberry may be attributed to the purposeful breeding and development of objective-specific cultivars. During the last eighty years, breeding work on strawberry has been conducted by several research institutions in the world. Many superior cultivars, possessing desired characters, have been developed. Among various public agencies, the New York Agricultural Experiment Station, Geneva, was the first to initiate the breeding work on strawberry in the United States during 1889 and there the emphasis was given on breeding for winter hardiness. Later, many other institutions also started breeding work on strawberry. Presently, public-supported breeding programmes are also active in Canada, South Africa, Europe, Asia, and the USSA. In India, breeding work was started at IARI, Regional Station, Katrain, Kulla Valley (HP), in the early sixties.

Modern Breeding

Major breeding objectives : The objectives of breeding programmes are similar but their relative importance differs among its breeders, primarily because of different conditions. For example, a disease or insect in a particular locality may be a serious problem there but not in other areas. Therefore, a breeder has to give more emphasis on developing a disease or insect resistant variety for that particular locality but for breeder of the other locality, it may not be of any high priority. In general, the commonly desired plant characters in breeding strawberries are yield, vigour, fruiting habit (long day, short day, or day neutral), ripening time, winter hardiness, runnering capacity, tolerance to high temperature, concentrated ripening, disease resistance (root diseases, foliage diseases, virus tolerance) and pest resistance (aphids, weevils and mites). Out of these, yield, vigour and fruiting habits are of primary importance, while others are considered as per local importance. The characteristics of fruits included are size, symmetry, shape, colour, firmness, flavour, ease of capping, vitamin C content, TSS, acidity and resistance to rot. Of these, size, firmness, and flavour are always important, while others are of variable concern, depending upon the locality and the major objective of the breeder.

Techniques of breeding

Floral biology : Usually three types of flowers like pistillate; staminate and hermaphrodite are found in strawberry. But sometimes, plants may be poly-ando-dioecious, with the first few flowers on predominately pistillate plants producing fruits. However, most of the present-day cultivars bear hermaphrodite flowers and therefore require emasculation for cross-pollination. Its flower has many pistils, each having its own style and stigma, attached to receptacle and on fertilization, develop into a fleshy fruit. The true fruits of strawberry are nutlets (achenes), containing one seed in each. Each flower may have 50 to 400 pistils that develop seeds. Usually, primary flowers have the largest number of pistils, followed by secondary and tertiary ones. The flower has five petals, one style and several stamens.

Pollination, seed collection and raising seedlings : In emasculation, all the flower parts are removed so that the pistil is fully exposed for pollination. Emasculation should be done within 1 to 3 days before anthesis to prevent selfing. The emasculated flowers must be protected from foreign pollen grains by bagging. Pollen grains of desired cultivar are collected 1 to 2 days before anthesis and kept in vials to dehisce. However, these can be stored at 4° C under low humidity for future use. Pollen grains are transferred to stigmas with a small camel's hairbrush. After pollination, fruits ripen in about 25 to 30 days under normal environmental conditions. Seeds may be collected from the pulp and are usually sown in germinating medium like sand, sterilized soil or shredded sphagnum moss in the seed trays. Since young seedlings are susceptible to damping off, they should be handled carefully during their early growth. These are transplanted into soil in flat pans or pots after 6 to 8 weeks in hill or matted row system of planting. For screening of seedlings against diseases, artificial inoculation with disease causing pathogens have been perfected. Similarly, for measuring resistance of seedlings to insect-pest in field, a greenhouse screening is done. Further, tests for virus tolerance usually require the planting or maintenance of the seedling in the field for at least two or more years in areas where virus and vectors are present in abundance. The selected seedlings are then evaluated subjectively or objectively in single or replicated plots. The most frequently used objective measurements are yield, fruit size, fruit weight and fruit firmness.

Breeding systems

Outbreeding and inbreeding : The outcrop method of breeding has been used mostly for the improvement of strawberries. In this system, most characters segregate in each generation and selection pressure must be exerted simultaneously on each character. The seedlings with desirable combination of characters are selected with the expectation that the new combination will be superior in characters to one, other or both the parents. In general, wide parentage base is necessary in outbreeding to avoid the chances of inbreeding, which often results in loss of vigor and reduction in the yield. Inbreeding has been used occasionally as a method for evolving new cultivars in strawberry.

Interspecific hybridization : All the octoploid *Fragaria* species are inter-fertile. *Fragaria chiloensis* and *F. virginiana* have been used extensively in the improvement of cultivated strawberry. The octoploid species of strawberry have many desirable characters and therefore these have been used widely for the improvement.

Strawberry breeders have exploited the winter hardiness of *F. virginiana* ssp. *glauca*, resistance of *F. chiloensis* to red stele, strawberry aphid, two spotted mite; resistance of *F. chiloensis* and *F. virginiana* to Verticillium wilt; virus tolerance of *F. chiloensis*, the day neutral character of *F. virginiana* ssp. *glauca*; large fruit size and firmness of *F. chiloensis*. The incorporation of all the desired characters into the cultivated strawberry is not an easy task because some undesirable characters like small fruit size, partial fertility, soft flesh, black seeds, susceptibility to leaf diseases and dull fruit colour are also inherited to the progeny. Thus, a native octoploid species should be used for breeding programme for raising at least 3 to 4 generations by outbreeding or back crosses to obtain a progeny of desired characteristics. Moreover, it requires a large population to select a seedling with desired characters from the segregated population.

Intergeneric hybridization : Various breeders have tried the intergeneric hybridization in strawberry but very low success has been achieved. Hybridization between *Fragaria* and its related genus *Potentilla* has been tried but most hybrids died at an early age or were sterile. Some scientists have made several attempts to cross *F. vesca* with different species of *Potentilla* but the resultant hybrids do not survive and die in early stages of their growth.

Inheritance of characters

Most of the plant and fruit characters of cultivated strawberries are inherited quantitatively. Additive, dominance and epistatic components may act in the expression of a particular character. It usually creates complex breeding problems when maximum expression of a combination of characters is desired. The inheritance pattern of some characters of strawberry is described briefly hereunder:

Yield : Yield is a component of many factors like number and size of fruits, plant health, hardiness, disease resistance of plant and crown number per plant and row. Fruit size differs within a fruit cluster, primary berry being the largest and others relatively smaller. Fruit size is inherited quantitatively. The heritability estimates for other yield components, like berry weight, berries/flowers, yield/flower stem, flower stems, are controlled by additive, dominant or epistatic genes.

Disease resistance : Inheritance of resistance for most of the diseases of strawberry is quantitative and additive. Breeding for resistance to Ramularia leaf spot is generally restricted to the races that occur locally. Therefore, breeding for resistance has been on a regional basis and cultivars that are resistant in one locality may be susceptible in the other. No cultivar has been found to be immune, although Dabreak is most resistant to Ramularia leaf spot (Table 8). Other cultivars showing regional resistance to leaf-spot are Albritton, Elista, Hood, Howard 17, Lassen, Laleglow, Surecrop and Tangi. Powdery mildew is widely prevalent on strawberry leaves and may infect flowers and fruits. The inheritance of powdery mildew resistance has not been fully understood, however, it may be partly due to the thickness of cuticle. Tioga and Brighton are resistant cultivars to powdery mildew (Table 8).

Verticillium wilt is a widespread disease of strawberry plants grown in soil previously occupied by solanaceous vegetable crops. Resistance to Verticillium wilt is partially dominant and is inherited quantitatively. None of the cultivars has been found to be immune, but Allstar, Catskill, Delite, Earliglow, Gem, Guardian, Hecker, Hood, Sierra, Tribute, Tristar and Totem are resistant to Verticillium wilt (Table 8).

Red stele is a destructive disease of strawberry under cool, moist condition. Original source of its infection is unknown but it is believed that *Potentilla* serves as the natural host. Red stele has many races and it is difficult to develop a cultivar resistant to all the races. Resistance to red stele is partially dominant and quantitatively inherited. Gorella, Juspa, Lavo, Tamella, Templer, Guardian, Hood, Midway, Redchief, Sparkle, Surecrop, and Sunrise are resistant to one or the other races of red stele (Table 8).

Anthracnose attacks the crowns, petioles and runners of strawberry under warm, humid conditions. The fungus has many races and thus the cultivars resistant in one location may be susceptible at another. Botrytis fruit rot is the most destructive disease of strawberry in field and storage. Two separate modes of inheritance appear to confer resistance to pre- and post-harvest rot, although resistance to both these stages has been combined in Earliglow (Table 8). However, the exact nature of resistance has not been clear but it appears to be quantitative and additive. In general, fruit firmness may play a vital role against grey-mould resistance.

Virus tolerance : Viral diseases cause heavy losses to strawberry where a large population of aphid or nematode is found. In breeding for virus tolerance, plants are either screened in fields or by transmission tests in greenhouses. High degrees of tolerance in several clones of *F. chiloensis* have been reported and there exist wide differences in virus tolerance among cultivars. However, Cheam, Northwest, Shuksan and Totem are tolerant cultivars to most of the viruses (Table 8).

Pest resistance : Inheritance pattern for resistance has been studied for a few insect-pests in strawberry. The resistance to two-spotted mites is highly heritable and quantitative. Lassen, Cardinal, Comet, Blackmore and Muir are known cultivars having resistance to two spotted mite (Table 7). The inheritance of resistance to strawberry aphid is still emerging but it appears to be controlled by more than a single locus, with additive gene action. De Norte and Yaquina clones of *F. chiloensis* are resistant. Similarly, Benton is also a resistant cultivar to aphid than Totem, Hood and Tyee.

Fruiting habit

The strawberry cultivars have been classified into three groups on the basis of bearing habit. They are (i) single cropper or June bearer, which is a characteristic of most of cultivars, (ii) ever-bearer, fruiting 2 to 3 times in a season, and (iii) day neutrals.

Many workers have studied the breeding behaviour of ever-bearing strawberries. It has been postulated that complimentary dominant genes governed the inheritance in ever-bearing strawberries and they segregate in an octaploid manner. Further, four recessive genes govern the expression of the ever-bearing character in strawberry. Similarly, ever-bearing fruiting habit in diploids is governed by a single recessive gene, which also acts independently as a single recessive gene for non-runnering capacity. The day-neutral character is regulated by a single dominant allele.

Winter hardiness : Winter injury is a serious problem in many strawberry growing areas. Strawberry cultivars differ invariably in their hardiness to winter injury. The inheritance of winter hardiness in strawberry is partially dominant over non-hardiness. Usually, the clones of *F. virginiana* ssp *glauca* have very high degree of winter hardiness. The European cultivars are also more winter hardy because they have large ancestral contributions from *F. virginiana*. Earlidawn and Howard 17 are frost tolerant cultivars (Table 9).

High temperature tolerance : Clones of *F. virginiana* show a high degree of resistance to high temperature (30° C) and humid heat than *F. chiloensis*. Blackmore, Missionary, Sparkle and Raritan cultivars can resist high temperature (Table 9). Similarly, Pusa Early Dwarf, Douglas, Fern, Sweet Charlie, Pajaro, Selva and Chandler grow well in north Indian plains, where temperature is quite high.

Flowering and ripening time : Time of flowering and ripening are closely related in strawberry. All these characters are inherited quantitatively.

Adaptability to mechanical harvesting : In the present-day breeding programmes, emphasis has been given on the development of cultivars suited for mechanical harvesting. The cultivars for mechanical harvesting should have high yields, uniform ripening, easy calyx removal, and fruits with long petiole and necks. Ease of calyx removal have shown additive gene action. Cultivars Pugget Beauty, Olympus, Totem, Redchief' are easy to cap. Saladin and

Silver Jubilee have long (>40 mm) pedicels, whereas Veeglow and Pugget Beauty have long-necked fruits, making them ideal for mechanical decapping.

Stolon production : The rate of stolon production is highly variable among different species and cultivars. The breeding behaviour of the runnering character is very complex. The non-runnering character is highly heritable in *Fragaria x ananassa,* which is closely related with the ever-bearing character but specific combining ability plays a prominent role in the stolon production.

Variegation : Variegation is an abnormal yellow-green mottling of strawberry leaves. It may occur in early growth of seedlings or may not occur until plants become old. Most of the researchers largely agree that transmission of this character to progeny does not fit Mendelian pattern and that cytoplasmic control accounts for it. Climax, Blakemore, Midland and Vermilion cultivars are relatively free from variegation (Table 8).

Rest period : Strawberry also shows dormancy and requires some amount of chilling hours to overcome it. Cultivars differ widely in their chilling requirements. Cultivars adapted to warm localities like Chandler, Douglas, Dover, Pajaro, Selva, Tioga, Mission and Festival have short rest period and can grow well in short days of winter (Table 9). In most cultivars of *F. chiloensis,* ancestory has been important in developing low-chilling cultivars. Similarly, *F. virginiana* ssp. *glauca* produce flowers without chilling period at lower elevations.

Edaphic adaptations : The *Fragaria* species differ widely in respect to their soil requirement. For example, *F. chiloensis* grows better on much sandy and well-drained soils than *F. virginiana* or *F. vesca.* The *F. chiloensis, F. orientalis, F. vesca* ssp *americana* and *F. virginiana* ssp *glauca* are drought tolerant. Similarly, *F. chiloensis* can tolerate very high salt levels (> 1200 ppm).

The cultivated strawberries appear to differ in their relative sensitivity to soil reaction. For example, British Sovereign can grow in delta soils with 4.5 to 5.0 pH, while Marshall cannot grow. In general, wild

Table 9 : A list of strawberry cultivars resistant to adverse soil and climatic conditions

Sl. No.	Condition	Cultivars
1.	Alkalinity tolerance	Solana
2.	Low temperature tolerance	Ogallala, Sparta, Jubilee, Gem, Senga Sengana etc.
3.	Resistance to frost	Earlidawn and Howard 17
4.	High temperature tolerance	Missionary, Blakemore, Florida Ninety, Sparkle, Raritan, Klondike, Headliner, Torrey, Lassen etc.
5.	Resistance to drought	Blakemore, Marshall, Surecrop etc.
6.	Low chilling requirement	Dabreak, Missionary, Blakemore, Florida Ninety, Klondike, Headliner, Torrey, Frenso, Tioga, Kloemore, Chandler, Douglas, Selva, Pajaro, Dover, Fern, Festival, Lassen etc.

strawberries grow vigorously on acid, neutral and alkaline soils. Some strawberry cultivars are resistant to one or the other adverse edaphic and climatic conditions (Table 9). Similarly, some cultivars have many desirable characters and can be used in future breeding programmes. A list of some of the most peculiar examples is given in Table 10.

Some achievements

Crop improvement is an endless process. Breeders provide solution to the present day problems by way of developing new cultivars but new problems come and breeders start working on that. Much has been achieved in strawberry since 1920 and much is to be achieved yet. Howard 17 was the premier variety of strawberry in 1900 having many desirable characters and later it was used in many breeding programmes to incorporate some of its desirable characters in its progeny. Thirty-nine varieties of strawberries have Howard 17 as one of the parents, whereas 87 varieties have Howard 17 in their

Table 10 : Strawberry cultivars with some desirable characters

Sl. No.	Desirable character	Name of the cultivars
1	Suitability for mechanical harvesting	Gorella, Pugget Beauty, Olympus, Totem, Redchief, Mimak, Primek etc.
2.	Easy in capping	Jucunda, Juspa, Gorella, Senga Sengana, Abundance, Tenneese Beauty, Fresno, Tioga, Torrey etc.
3.	Good processing qualities	Midland, Earlidawn, Tioga, Torrey, Morella, Marshall, Sparkle, Catskill, Abundance, Senga Sengana etc.
4.	Suitability for freezing	Midland, Earlidawn, Fairfax, Albritton, Hapil, Earliglow, Northwest, Blakemore, Redcoat, Dixieland, Klondike, Fletcher, Senga Sengana etc.
5.	Excellent flavour	Midland, Earlidawn, Fairfax, Marshall, Albritton, Cambridge Vigour, Solana, Cambridge Favourite, Talisman, Earliglow, Northwest, Korona, Dorsett, Fletcher and Senga Precosa etc.
6.	High ascorbic acid content	Catskill, Tenneese Beauty, Marshall, Sparkle, Northwest etc.
7.	Attractive red flesh	Cardinal, Cornvallis, Earlibelle, Earlidawn, Rainier, Shuksan, Totem, Tyee, Vantage, Veegem, Vibrant etc.
8.	Firm flesh	Blakemore, Tenneese Shipper, Albritton, Apollo, Cardinal, Cornwallis, Korona, Fresno, Gorella, Selva, Pajaro, Tioga, Hapil, Hoilday, Senga Sengana etc.
9.	Large fruits	Belrubi, Shasta, Solana, Florida Ninety, Catskill, Douglas, Gorella, Guardian, Jewel, Kent, Pajaro, Robinson, Ceylon, Earlidawn, Titan, Totem, Vesper etc.
10.	Suitability for greenhouse cultivation	Aliso, Rosella, Honoyae, Gorella, Belrubi, Elsanta, Brighton, Fern, Selva, Fresno Lassen, Douglas, Chandler, Redgauntlet etc.
11.	Good glossyness	Vesper, Earlidawn, Albritton, Sparkle, Tioga, Redcoat, Empire, Senga Precosa, Redgauntlet etc.

ancestry. Some of the commercial varieties like Beever, Blackmore, Catskill, Earlimore, Fairmore, Midland, Red Gold, Robinson, Tennesse Beauty, Tennesse Supreme and Valentine have Howard 17 as one of the parents. Similarly, Albritton, Cambridge Favourite, Dixieland, Fletcher, Headliner, Kloemore, Midway, Pugget Beauty, Redcoat, Redglow, Redgauntlet, Shasta, Sierra, Surecrop, Tioga, Torrey and Vermilion have Howard 17 in their ancestry.

The rapidity with which the new cultivars are replacing the older ones is a testimonial to the success of the present-day strawberry breeders. Most recent introductions may replace the older cultivars more rapidly than in the past but more work is still needed on the inheritance of characters, breeding behaviour of parents and incorporation of new characteristics to hybrid seedlings. Present-day breeding has resulted in the development of many objective-specific cultivars but still we need an ideal variety of strawberry having satisfactory combination of all desirable plant and fruit characters. No doubt, it is a very difficult job but is under the accessible limits of modern strawberry breeders. In future, one of the greatest challenges will be to breed cultivar resistant to all races of red stele root-rot, Verticillium wilt and foliage diseases (leaf-spot, scorch, blight etc.), mildew, anthracnose, viral infections, botrytis rot, mites and nematodes. For this, breeders have to search the new sources of resistance and have to employ new technologies like RFLP'S and RAPD'S for identification and tagging of genes of interest. Similarly, developing of genomic maps and use of molecular technologies will be highly useful for the strawberry breeders.

Tribute, Tristar, Chandler, Sweet Charlie, Fern, Pajaro, Douglas, Jewel, Capitola, Camarosa, Bountiful, Blomidon, Totem, Korona and Senga Sengana are some important varieties developed during the last two decades at different research institutes of world.

In India, no systematic breeding work has been done in strawberry. However, major work has been done on the introduction and evaluation. The I.A.R.I., New Delhi, was the first institution in India to start work on strawberry. This institute released Pusa Early Dwarf and Katrain Sweet in late sixties. Similarly, the scientists of Himachal Agricultural University, Palampur, have recommended Tioga and Torrey for commercial cultivation in mid-hills of Himachal Pradesh. In Uttar Pradesh, Redcoat has been reported to be superior. Similarly, Chandler, Pajaro, Douglas and Fern etc. have been found to perform better under sub-tropical and tropical climates of India. Recently, IARI Regional Horticultural Research Station, Shimla, has recommended Shimla Delicious and Jutog Selection for cultivation in mid-hills of Himachal Pradesh.

An ideal variety

An ideal strawberry variety should have large fruits, tough skin, better processing qualities and better quality attributes. Some of the cultivars having such desirable attributes are given below:

Large fruited varieties : Large fruit size is preferred in strawberry. Some large-fruited cultivars are: Catskill, Douglas, Gorella, Jewel, Kent, Pajaro, Titan, Totem and Vesper (Table 11). Similarly, in India, Belrubi, Chandler, Fern, Senga Sengana, Selva and Tioga have been found to be large fruited cultivars and can be utilised in future breeding programmes.

Flesh firmness : Berries with tough skin and flesh are considered best for post-harvest handling and value-addition. Apollo, Holiday, Korona, Blakemore, Pajaro, Selva and Tioga have notable firm flesh (Table 10).

Better processing qualities : Poland and the United States process strawberries into various value-added products. The fruits for freezing purpose should have uniform flesh colour, firmness, high flavour with tartness, and ease of capping or plugging. Moreover, frozen berries should retain their flavour after processing. Fairfax is considered as the best cultivar having very good flavour. Similarly, Albritton, Earliglow and Korona have also fresh market flavour (Table 10). Some cultivars have attractive red flesh colour and flavour and

Table 11 : Cost and returns analysis of strawberry cultivation on per hectare basis

1.	Cost of plants (50,000/ha) @ Rs. 2/ plant	Rs.1,00,000/-
2.	Land preparation	
	Ploughing (One tractor @ Rs. 1000/hr for 3 hours)	Rs. 3,000/-
	Cost of FYM and insecticide (50 tonnes FYM @ Rs.300/t)	Rs. 15,000/-
	Furrow/bed preparation (50 labourers @ Rs. 90/day)	Rs. 4,500/-
3.	Planting cost	
	a. Fungicide treatment (cost of fungicide + 2 labourers @ Rs. 90/day)	Rs.500/-
	b. Planting (50 labourers @ Rs 90/- per day)	Rs. 4,500/-
	c. Watering with hand fountain (5 labourers @ Rs. 90/- per day)	Rs. 500/-
4.	Installation of micro-irrigation unit	Rs.60,000/-
5.	Intercultural operations	
	a. Fertilizer application (Cost of fertilizers + application charges)	Rs. 10,000/-
	b. Mulching (cost of mulching material + labour cost)	Rs. 5,000/-
	c. Weeding (cost of weedicide + labour cost)	Rs. 1,000/-
	d. Insect/pest and disease control (cost of insecticides, fungicides + labour cost)	Rs. 5,000/-
	e. Plastic cover for 2-3 months	Rs. 10,000/-
	f. Nylon netting	Rs. 10,000/-
6.	Fruit harvesting (five labourers @ Rs. 90/- per day for 30 days)	Rs. 12,000/-
7.	Overhead expenses	Rs. 5,000/-
8.	Bank interest on loan or otherwise	Rs. 10,000/-
9.	Total expenditure	Rs. 2,61,000/-
10.	Income	
11.	a. Sale of fruits (8 tonnes @ Rs. 55/Kg)	Rs. 4,40,000/-
	b. Sale of runners(25,000 plants @ Rs. 2/plant)	Rs. 50,000/-
12.	Gross income	Rs. 4,90,000/-
13.	Net profit	Rs. 2,29,000/-

they also retain it after freezing. Cardinal, Cornwallis, Totem, Tyee, Vibrant and Veegem cultivars have attractive red flesh (Table 10).

Better quality attributes : In addition to fruit size, shape, colour and flavour, TSS, acidity and vitamin C contents are considered important quality attributes in strawberries.

Chapter 26

Cultivation Cost & Cultural Hints

Cost of Cultivation

Cost analysis is a very important part of every farming enterprise. The main aim of a grower is to get maximum possible returns from his land and the investment made by him on the business. Moreover, it becomes more imperative in present days because the land and other resources are becoming limited and costlier day-by-day. The profit of a farming business lies basically on efficient farming efficiency and the input costs. In crop production, various kinds of material and services are required, which are highly heterozygous in nature. But, in general, the production cost includes the cost of land, human labour, animal labour, machine power, building, input material cost etc. Production cost may also be grouped according to the cultural operations. The cost of strawberry production generally includes the cost involved in the land preparation, planting material, pre-planting treatment, manures and fertilizers, planting, mulching, irrigation, protection from insect-pests and diseases, protection from birds, picking, packing and transportation cost etc. Whatever may be the cost, in every instance, money invested for land, labour and equipment and input material is included in calculating the cost of cultivation. It is really a tough job to work out the cost of cultivation for a particular locality because of changing rates and availability of labour, cost and planting of runners, cost of fertilizers, insecticides and fungicides etc., change from time to time and place to place. However, the cost of strawberry cultivation is usually higher in the hills compared to plains. By following scientific management practices, a grower can earn as high as 2.50 lakhs from one hectare strawberry field. The rough estimate of cost of cultivation of strawberry is given in Table 11.

Some important cultural hints

For successful cultivation of strawberry, one should take every care to follow the following hints:

1. Start the strawberry cultivation in a locality where some plantings are already existing. It will help in reducing the marketing and transportation problems by joining hands together. Moreover, the joint cultivation will also reduce the severe damage caused by the birds.
2. Get your soil and irrigation water tested before starting commercial strawberry cultivation.
3. Try to start its cultivation in a locality well connected by roads.
4. The locality should have plenty of sweet water or have facilities for irrigation water.
5. Plenty of labour should be available in the locality, where strawberry plantings have been planned.
6. It is a highly perishable fruit so its planting should be started in areas located near a town or a city and near to a canning or processing units.
7. Always try to produce plants of yours or procure plants either from Government approved nurseries/Agriculture University or from ICAR research institutes/stations. If possible use tissue cultured plants for better profits/returns.
8. Planting should always be done on raised beds or in furrows.
9. Use healthy and well-rooted runners for planting.
10. Initially, try 3-4 varieties in your locality. Never try a single variety, as it may be a complete failure.
11. Do planting at proper time. The right time for planting in hills is August-September, while late October planting gives good success in plains.
12. Strawberry is a shallow rooted crop. Irrigation at right time re-

sults in better quality crop. It requires less but frequent water supply.

13. While irrigating strawberries, try to avoid the direct contact of water with plants. Always irrigate furrows not the plants.
14. For efficient use of water and better results, establish sprinkler or drip irrigation network.
15. Mulching either with paddy straw or hay or black polyethylene is absolutely necessary. In hills, pine needles are cheap and the best mulching material. In general, locally available material should be used as a mulch.
16. Monitor the insect-pests and diseases in proper time. Use recommended doses of insecticides and fungicides, whenever necessary.
17. Birds, especially sparrows, peacocks and parrots cause heavy damage to strawberries, so protection from birds by installing reflecting ribbons or bird-scarer is must. If possible, spread nylon net over the entire field.
18. Harvesting should be done at right time. As far as possible, avoid delayed picking as it may result in complete rotting of the fruits. Usually, the fruits are ready for harvesting when 2/3rd portion of the fruit has attained red colour.
19. Harvesting should preferably be done either in the morning or late evening hours.
20. Picking may be done every day for local markets and thrice a week for distant markets.

Selected References

Albregts, E.E. and Howard, C.M. 1978. Evaluation of plant density on strawberry plant growth and fruiting response. *Proc.Fla. Sta. Hort. Soc.*, **91** : 298-299.

Albregts, E.E; Howard, C.M; Chandler, C.K. and Martin, E.G. 1990. Fruiting response of strawberry as affected by rates and sources of controlled release of N fertilizer and irrigation method. *Proc. Soil Crop Sci. Soc. Fla.*, **49** : 46-49.

Anonymous, 1997. Package and practices for fruit cultivation in Himachal Pradesh, pp 48-49.

Awasthi, R.P. and Badiyala, S.D. 1983. Spacing and performance studies in some strawberry cultivars. *Indian J. Hort.*, **40 (1 & 2)** : 29-34.

Badiyala, S.D. and Joolka, N.K. 1983. Effect of different spacings on the performance of strawberry cv. Tioga. *Haryana J. Hort.*, **12(3-4)** : 165-167.

Badiyala, S.D. and Bhutani, V.P. 1990. Effect of planting dates and spacing on yield and quality of strawberry cv. Tioga. *South Indian Hort.*, **38(6)** : 295-296.

Bagnara, D. and Vincent, C. 1988. The role of insect pollination and plant genotype in strawberry fruit set and maturity. *J. Hort. Sci.*, **63(1)** : 69-75.

Brown, S.M; Whitwell, T. and Pounders, C.T. 1984. Weed control systems in strawberries. *Proc. Southern Weed Sci. Soc.*, 37th Annual Meeting, pp.139.

Catcliffe, J.A. and Blatt, C.R. 1984. Effects of N, P, K, B, and lime on strawberry yields. *Canadian J. Pl. Sci.*, 64(4): 945-949.

Chandler, C.K.; and Albregts, E.E; Howard, C.M. and Brecht, J.K. 1997. Sweet Charlie strawberry. *Hort. Science*, **32(6) :** 1132-1133.

Childers, N.F; Morris, J.R. and Sibbett, G.S. 1995. Modern Fruit

Science, Hort. Publication, Gainesville, Florida, USA.

Collins, W.B. 1966. Effect of winter mulches on strawberry plants. *Proc. Am. Soc. Hort. Sci.,* **89** : 131-135.

Darrow, G.M. 1966. The strawberry History, Breeding and Physiology. Holt, Rinehart and Winston, New York, Chapter, 20.

Daubeny, H. A. 1994. Strawberries. *Hort. Science,* **29(9)** : 960-964.

Daubeny, H. A. and Kempler, C. 1997. Nanaimo strawberries. *Hort. Science,* **32(7)** : 1293-1294.

Denisen, E.L. 1983. Mars strawberries. *HortScience,* **18(3)** : 498.

Dennis, F.G. and Bennet, H. O. 1969. Effect of gibberrellic acid and deflowering on runner and inflorescence development in an early bearing strawberry. *J. Amer. Soc. Hort. Sci.,* **94** : 558-560.

Draper, A.D; Gallette, G.J. and Swartz, H . J. 1981. Tribute and Tristar everbearing strawberries. *Hort. Science,* **16(6)** : 794-795.

Dwivedi, M.P. 1987. Effect of photoperiod and growth regulators on vegetative growth, flowering and yield of strawberry. *Ph. D. thesis,* Dr. Y.S. Parmar University of Hort. and Forestry, Solan, India.

F.A.O. production year book (1998)

Feng, Y.C. 1998. Three cultural systems for strawberry growing and the key techniques. *South China Fruits,* **27(5)** : 47.

Galletta, G.J. and Bringhurst, R.S. 1990. Strawberry management. *In* : Small Fruit Crop Management, Galletta, G.J. and Mimelrick (Eds.). Pretince Hall. Englewood Cliffs, N.J.

Galletta, G.J; Draper, A.D. and Swartz, H . J. 1980. Scott strawberry. *Hort. Science,* **15(4)** : 541-542.

Galletta, G.J; Draper, A.D. and Maas, J.L. 1987. Lester strawberry. *Hort. Science,* **22(2)** : 321-323.

Galletta, G.J; Maas, J.L; Enns, J.M; Scheerens, J.C; Rouse, R. and Heflebower, R.F. 1996. Primetime strawberry. *Hort. Science,* 31(6) : 1038-1042.

Galletta, G.J; Maas, J.L; Enns, J.M; Scheerens, J.C; Rouse, R. and

Heflebower, R.F. 1996. Latestar strawberry. *HortScience,* **31(7)** : 1238-1242.

Garren, R. 1980. Causes of misshaped strawberries. In: The strawberry, Childers, N.F. (Ed.), Horticultural Publications, Gainsville, Fl. Pp. 326-335.

Gast, K.L.B. and Pollard, J.E. 1991. Row covers enhance reproductive and vegetative yield componenets in strawberries. *HortScience,* **26** : 1467-1469.

Goodman, R.D. and Oldroyd, B.P. 1988. Honey bee pollination of strawberries (*Fragaria x ananassa* Dusch.). *Aust. J. Expt. Agric.*, **28** : 435-438.

Guttridge, C.G. 1958. The effects of winter chilling on the subsequent growth and development of the cultivated strawberry plant. *J. Hort. Sci.,* **33** : 119-122.

Guttridge, C.G. 1969. Fragaria. In : The induction of flowering, Evans, L.T. (Ed), MacMillan, Inc. Melbourne, Chapter 10.

Guttridge, C.G. 1985. *Fragaria x ananassa.* In : CRC hand book of flowering, Vol. 3, Halevy, A. H . (Ed), CRC Press Boca raton, FL, Chapter 1.

Hartman, H . T. 1947. Some effects of temperature and photoperiod on flower formation and runner production in the strawberry. *Plant Physiol.,* **22** : 407-410.

Hertz, L.B. 1979. The effectiveness of sudan grass straw and polyethylene mulches on growth and yield of Trumpeter strawberry. *HortScience,* 14(3): 236-238.

Heide, O.M. 1969. Photoperiod and temperature interactions in growth and flowering of strawberry. *Physiol. Plant.,* **40** : 21-25.

Howard, C.M. and Albregts, E.E. 1980. Dover strawberries. *HortScience,* **15(4)**: 540.

Hughes, H. M; Duggan, J.B. and Banwell, M.G. 1969. Strawberry bulletin No 95, HMSO 10, 6th edition, Ministry of Agri. Fish Food, UK.

Kapur, O.C; Sharma, S.K; Masand, S.S. and Chakor, I.S. 1991. Effects of phosphorus and irrigation management on water use and yield of strawberry in H . P. *Himachal J. agri. Res.,* **17(1-2)** : 154-157.

Korenberg, H. G. 1959. Poor fruit setting in strawberries I. Causes of poor fruit set in strawberry in general. *Euphytica,* **8** : 47-49.

Kumar, J; Rana, S.S; Verma, H . S. and Jindal, K.K. 1996. Effect of various growth regulators on growth, yield and fruit quality of strawberry cv. Tioga. *Haryana J. Hort. Sci.,* 25(4) : 168-171.

Larson, K.D. and Shaw, D.V.1995. Relative performance of strawberry genotypes on fumigated and non-fumigated soils. *J. Amer. Soc. Hort. Sci.,* **120 (2)** : 274-277.

Lieten, F. 1989. Strawberry : albisinm, a new physiological disorder. *Fruitteelt,* **2(9)** : 39-41.

Lieten, F. and Marcelle, R. 1993. Relationship between fruit mineral content and albinism disorder in Elsanta strawberry plants. *Acta Hort.,* **348** : 294-298.

Locascio, S.J; Myers, J.M. and Martin, F.G. 1977. Frequency and rate of fertilization with trickle irrigation for strawberries. *J. Amer. Soc. Hort. Sci.,* **102**: 456-458.

Mitra, S.K. (1991). Strawberry *In:* Temperate fruits. Naya Prokash Publishers, calcutta, India.

Miura, H; Imqda, S. and Yabuchi, S. 1990. Double sigmoid growth curve of strawberry fruit. *J. Japanese Soc. Hort. Sci.,* **59(3)** : 527-531.

Moore, P.P., Sjulin, T.M. and Shanks, C. H. 1995. Puget Reliance strawberry. *HortScience,* 30(7): 1468-1469.

Ohishi, T. 1999. Appropriate management of honeybee colonies for strawberry pollination. *Honeybee Sci.,* **20(1)** : 9-16.

Pandey, D; Phogat, K.P.S and Shukla, S.N. 1980. A note on the effect of different spacings on growth and yield of strawberry cv. Red Coat. *Progressive Hort.,* **12(3)** : 67-70.

Phogat, K.P.S; Pandey, D. and Shukla, S.N. 1980. Effect of different methods of planting on growth and yield of strawberry Cv. Red Coat. *Progressive Hort.,* **11(4)** : 31-33.

Rosati, P. 1993. Recent trends in strawberry production and research: An overview. *Acta Hort.,* **348 : 23-43.**

Schrevens, E; Lamberts, D. and Lettani, L. 1989. Intensive hydrophonical cropping system for strawberries in green houses. *Acta Hort.*, **265** : 275-282.

Scott, D. H . and Lawerence, F.J. 1975. Strawberries. *In* : Advances in Fruit Breeding. Janick, J and Moore, J.N. (Eds.). Purdue Univ. Press., West Latayette.

Sharma, O.P. and Phogat, K.P.S. 1983. Effect of plant growth regulators on vegetative growth, yield and quality of strawberry (*Fragaria spp*). *Progressive Hort.*, **15(1)** : 64-68.

Sharma, R.L. and Badiyala, S.D. 1980. A study on the performance of some strawberry (*Fragaria x ananasa*) cultivars in the mid hills of north India. *Progressive Hort.*, **12(2)** : 17-24.

Sharma, R.R. (1999). Strawbeerry ki kheti. *Prashar Doot.*, **3(2)** : 17-20.

Sharma, R.R. (2001). Strawberry : Improved production technology. Ministry of Agriculture, Govt. of India, New Delhi.

Sharma, R.R., Sharma, V.P. and Meena, Y.R. (2001). Strawberry cultivation : A profitable business. *Intensive Agri.*, **38(1)** : 22-23.

Sharma, R.R. and Singh, S.K. 1999. Strawberry cultivation - a highly remunerative farming enterprise. *Agro India*, **3(2)** : 29-31.

Sharma, V.P. and Sharma, R.R. 2000. The strawberry. To be published by ICAR, New Delhi.

Sharma, V.P. and Singh, Ranjit . 1990. Growth and fruiting behaviour of strawberry (*Fragaria spp.*) as affected by cloching and gibberellic acid treatments. *Proc. 11th Intl. Congress on use of plastic in Agric*, New Delhi, pp. 147-149.

Shoemaker, J.S. 1975. Strawberries. *In* : Small Fruit Culture, The AVI Publishing Company, Westport Connectiout, USA.

Singh, R.P; Srivastava, R.P. and Phogat, K.P.S. 1983. Studies on foliar nutrition of strawberry, *Punjab Hort. J.*, **23(1)** : 64-68.

Singh, S.J. 1990. Diseases of strawberry. *In* : Advances in diseases of fruit crops in India. Kalyani Publishers, Ludhiana,India, pp. 397-401.

Stahler, M.M; Lawrence, F.J; Martin, L.W; Moore, P.P; Daubeny, H .

A; Sheets, W.A. and Verseveld, G.W. 1995. Redcrest strawberry. *HortScience,* **30(3)**: 635-636.

Takeda, F. and Papadopoulos, A.P. 1999. Strawberry production in soilless culture systems. *Acta Hort.,* **481 : 289-295.**

Tosi, T. 1986. Some aspects of strawberry fertigation. *Acta Hort.,* **176** : 99-106.

Verma, S.K. and Phogat, K.P.S. 1995. Effect of pollination by honey bees on fruit setting and yield of strawberry (*Fragaria sp.*) under Himalayan valley conditions. *Progressive Hort.,* **27(1-2)** : 104-108.

Vince-Prune, D. and Guttridge, C.G. 1973. Floral initiation in starwberry : Special evidence for the regulation of flowering by long day inhibition. *Planta,* **110** : 165-168.

Westwood, M.N. 1993. Temperate zone pomology. W. H . Freeman and Co., San Francisco.

Wilhelm, S. and Nelson, R.D. 1981. Fungal diseases of strawberry. *In* : The strawberry, Childers, N.F. (Ed.) Horticultural Publications, Gainsville, Fl.

Yoshida, Y; Ohi, M. and Fujimoto, K. 1991. Fruit malformation, size and yield in relation to nitrogen nutrition and nursery plants in large fruited strawberry (*Fragaria x ananassa*). J. *Japanese Soc. Hort. Sci.,* **59(4)** : 727-735.

Index